TRANSPORT, DEMAND MANAGEMENT
AND SOCIAL INCLUSION

T0359021

For Maureen and Dilip, for everything

Transport, Demand Management and Social Inclusion

The Need for Ethnic Perspectives

FIONA RAJÉ,
Transport Studies Unit,
University of Oxford

with assistance from

MARGARET GRIECO
Napier University and Cornell University

JULIAN HINE
University of Ulster

JOHN PRESTON
University of Oxford

Routledge
Taylor & Francis Group

LONDON AND NEW YORK

First published 2004 by Ashgate Publishing

2 Park Square, Milton Park, Abingdon, Oxon OX14 4RN
711 Third Avenue, New York, NY 10017, USA

Routledge is an imprint of the Taylor & Francis Group, an informa business

First issued in paperback 2017

British Library Cataloguing in Publication Data
Transport, demand management and social inclusion : the
   need for ethnic perspectives. - (Transport and society)
   1. Transportation - Social aspects - Great Britain
   2. Transportation - Social aspects - Great Britain - Case
   studies 3. Transportation and state - Great Britain
   4. Transportation and state - Great Britain - Case studies
   5. Minorities - Great Britain - Economic conditions
   6. Minorities - Great Britain - Economic conditions - Case
   studies 7. Minorities - Great Britain - Social conditions
   8. Minorities - Great Britain - Social conditions - Case
   studies 9. Marginality, Social - Great Britain 10. Great
   Britain - Ethnic relations - Economic aspects
   I. Rajé, Fiona
   388'.0941

Library of Congress Cataloging-in-Publication Data
Transport, demand management and social inclusion : the need for ethnic perspectives /
   Fiona Rajé.
      p. cm. -- (Transport and society)
   Includes bibliographical references and index.
   ISBN 0-7546-4045-0
      1. Transportation--Social aspects--Great Britain. 2. Marginality, Social--Great Britain.
   3. Transportation and state--Great Britain. I. Rajé, Fiona, 1966- II. Series.

   HE243.T6845 2004
   303.48'32'0941--dc22

                                                                              2004046331

ISBN 978-0-7546-4045-5 (hbk)
ISBN 978-1-138-25485-5 (pbk)

Contents

List of Tables

List of Figures

Chapter 1

Transport and Social Exclusion: A New British Policy Agenda

1.1 Introduction: capturing policy contexts and transporting excluded voices

I can't get the bus, only like one or two, with the pram. Like usually if I go to my friend's house, there is a bus and she lives quite far. But you can wait up to an hour for that bus and you can't get on it with a pram. So I have to go all the way up to town and then get on a bus which has space at the front of the bus for a pram and you can sit by your pram and the bus lowers.
(Black Caribbean Single Mother, 19 years old, Easton, Bristol)

I live in Montpelier, my daughter lives in Easton and I can't catch a bus to Easton. I have to go into town which is the complete other direction and then I have to catch a bus to Easton but it won't even go anywhere near where she lives. I have to get off at bottom of Stapleton Road and it's very expensive to have to catch 2 buses just to go a mile and a half to visit my daughter. And I've got another situation when my daughter was in labour and we had to go to St Michael's Hospital which is on the hill in Kingsdown and is one of the steepest hills in Bristol and it's quite a way and there's no bus service. So we had to walk with her in labour because we couldn't afford a taxi. The nearest bus is at the top of the hill, the 8 and you have to walk down. She was up 3 times that weekend and we could only afford the taxi once. It was really serious. They closed the ward and wouldn't let her stay even though she was in the beginning stages of labour and we had to go home with the worry that we couldn't afford any more taxis to get back there again. It was a nightmare.
(White Mother of Black Caribbean Single Mother, 40 years old, Montpelier, Bristol)

You find that in many UK cities the wards near the city centre were neglected for many years but now everyone wants to work, live near by and have leisure there too. So there is a push factor, the old communities get run down, land prices depreciate and a developer comes along and pushes up housing prices. Then people have to live further away and commute by car and have a problem with parking in the city centre and cause great congestion. I think successive governments have not really dealt with the issue of transport in the UK. John Prescott and his merrymen should have gone to Europe for 2 weeks and see what Germany, Holland, France, Sweden do and then make strategic decisions.
(Asian man, Outreach Worker, 35 years old, Easton, Bristol)

1

On my way to school, there's often too many rough people on the buses.
(Asian schoolboy, 14 years old, Easton, Bristol)

(In answer to the question: What journeys would you like to make by public transport and cannot?) Go to college. South Nottingham College is in West Bridgford which is like the next place to here. And it's a Trent Bridge bus. Not only that getting to Trent Bridge, you'd have to get a bus down to the bottom of the estate, get the Trent Bridge bus there, get to Trent Bridge cross over the road and get another one. That's 3 buses to go 10 minutes in the car. I was actually a volunteer helping with English classes for people who speak other languages and I had to give it up because it was 4 hrs a day to get there for a 2 hr lesson. It was ludicrous.
(White woman attending a young mother's group, Clifton, Nottingham)

These voices on transport and social exclusion in Britain are rarely heard directly: transport inconvenience, transport fear and simply the lack of transport are the common experience of much of low income Britain. Typically these voices are mediated through the looking glass of transport experts and policy gurus. The lived experience of public transport in modern Britain requires a more thorough recording. The objective of this book is to bring together the expression of these voices with an understanding of current changes in the transport policy arena and traffic management environments. The book begins with an overview of the current understanding of transport's relationship to social inclusion/ exclusion.

The interaction between social inclusion/exclusion and transport has not been adequately charted in Britain to date. However, the focus on the part poor transport links play in social deprivation and exclusion is rapidly strengthening as discussion of neighbourhood renewal and regeneration have become part of the policy agenda: one summary of the state of the field is provided by renewal.net:[1]

Poor transport links within deprived neighbourhoods, and between deprived neighbourhoods and the wider community, mean:

Individuals:
- can be cut off from jobs, education and training;
- may be unable to access cheap, fresh food;
- may only access health care in a crisis;
- are often unable to see friends and family or take part in other social activities;
- may experience crime or fear of crime walking to, waiting for, and travelling on public transport.

Communities may be isolated from services and opportunities in the wider economy. High levels of traffic and poor access can reduce the prospects for investment in towns and cities as well as making the local environment less pleasant.

Businesses: may suffer from lost customers and difficulty hiring employees.
(renewal.net, Poor transport links, 2002)

In the past few years, social inclusion/exclusion has emerged in transport-related discourse largely stimulated by policy agencies rather than by transport professionals. Until recently (Social Exclusion Unit, 2003), the coupling of the term social inclusion/exclusion with transport operated more strongly at the level of rhetoric rather than in the commissioning of substantial research into the relationship between social inclusion/exclusion and transport (Grieco et al., 2000). In the not so distant past, Kintrea and Atkinson (2001), writing about neighbourhoods and social exclusion, state that social exclusion had not been a matter of great concern in UK transport policy:

> ...the White Paper on Transport (DETR, 1998) concentrated mainly on themes such as congestion, car use and the environmental impact of transport, rather than the social aspects of transport provision.

Although the present policy framework around transport provision in relation to social inclusion/exclusion is largely underdeveloped, there has been some movement towards a greater understanding of the social aspects of transport in the research sphere. Social inclusion/exclusion has been entering the discourse through high level policy inputs such as the recent report by the Social Exclusion Unit (2003) in the Office of the Deputy Prime Minister which highlights the present failings of transport in respect of remedying social exclusionary processes. Despite the importance of this policy land mark, the Social Exclusion Unit report remains largely a consolidation of information on transport and social exclusion which was available previously albeit in a highly fragmented form and in a relatively weak professional field. This consolidation of materials represents a very important service, however, significant primary work on transport and social exclusion has yet to be undertaken. In a context where new primary work on transport and social exclusion is relatively scarce, this book reports on material gained in a study of transport and social exclusion in relation to traffic demand management measures (Road User Charging and the Work Place Parking Levy) for the British Department for Transport.[2] Over a 15-month period, the team from the Transport Studies Unit, University of Oxford undertook research into the social inclusion/exclusion impacts of Road User Charging (RUC) and Workplace Parking Levy (WPPL) focused, in particular, on gender, ethnicity and life cycle issues for the Department for Transport (DfT).[3] This built upon earlier research on gender (Hamilton et al., 1999), ethnicity (Beuret et al., 2000) and life cycle (for example: Turner and Pilling, 1999, on the young or Metz, 2000, on the elderly). This approach to the interface between transport management, ethnicity and social inclusion is a new innovation. However, new policies of traffic demand management raise old and new issues in respect of social equity in the transport domain and adjacent sectors.

Researching transport and social exclusion in the context of new proposed demand management measures in the cities of Bristol and Nottingham, and the surrounding policy literature, it became very evident that the transport policy professionals, with some notable exceptions, were very aware of the contributions that demand management measures could make to the improved efficiency of the cities but very low on the awareness of the public transport problems faced by many in the low income situation. Similarly, the transport disadvantaged public were very aware of high levels of public transport problems in low income situations but had a low awareness of the potential for transport efficiencies in new demand management measures and even in most cases a low awareness of the existence of demand management measures at all. This situation is one in which public consultation measures have clearly been insufficient to meet the needs of the audience in either direction: transport professionals are talking past the public in a brave new world of transport measures. Adequate channels for public consultation on the real nature of public transport problems have not been developed. The present policy situation is one in which two communities of interest – transport professionals and the user public – are talking past one another with a great deal of frustration being experienced on either side. The public acceptability of demand management and the 'public' efficiency of the public transport system require explicit policy attention. This book provides, based on two case study cities – Bristol and Nottingham – a preliminary foundation for such an approach. In addition, this particular book approaches this theme from the perspective of ethnicity – an area that has, thus far, had little discussion in the traditional transport literature.

1.2   Transport and social exclusion: the evidence on ethnicity

…despite the growing salience of the term 'social exclusion,' there was little consensus or agreement on its exact meaning or definition.
(Hine and Mitchell, 2001)

(social exclusion is) …a shorthand term for what can happen when people or areas suffer from a combination of linked problems such as unemployment, poor skills, low incomes, poor housing, high crime, bad health and family breakdown.
(Social Exclusion Unit, 2001a)

A key aim of the literature search was to develop a review of current research and writings on ethnicity and transport including visible religious minorities. This has been disappointing and has constantly demonstrated that most research (for example Modood 1997) into ethnic minority issues has neglected the influence of transport, especially in relation to other aspects of social policy such as health, welfare, housing and employment. To the extent that there is research, it has concentrated on crime, personal security and links with transport use. There is more information available in the American context, but this is specific to different cultures and not transferable to the British context. This dearth of research (which in some cases seems a glaring omission) certainly validates the decision of the Mobility Unit to commission this research. Throughout our visits to research institutes and university departments, the common reaction was consternation that transport issues had been forgotten.

(Extract from unpublished research on transport and ethnicity commissioned by the Department for Transport, Beuret et al. 2000)

The evidence on ethnicity in relation to travel, transport and social exclusion is unnecessarily weak: for example, despite a strong sociologicial literature on the maintenance of seclusion dynamics within certain categories of Asian immigrant households in the UK, there has been little attempt within travel behaviour analysis to explore this dynamic. Even within the framework of Household Activity approaches to travel behaviour within the UK, the resource tensions between the genders and the associated household travel dynamics are largely unexplored. Literature on social exclusion and transport (Sinclair and Sinclair, 2001) which is explicitly concerned with inequities in access and explores such issues often fails to discuss the ethnic dimension. Later in this section, we draw together the small body of materials which exist on ethnicity, transport and social exclusion – albeit the case that some of these provide indirect rather than direct evidence – however, to begin the discussion on transport and social exclusion in advance of exploring the ethnic dimension, some introduction to the available definitions and perspectives now operating in the field is necessary.

Weinberg and Ruano-Borbalan (1993) identified "the impossibility of agreeing the status of the 'excluded' by a single and unique criterion". Social exclusion is a highly contested term (Hills et al., 2001), despite the creation of bodies such as the Social Exclusion Unit of the Office of the Deputy Prime Minister and the ESRC's Centre for Analysis of Social Exclusion, based at the London School of Economics. The origins of the term are associated with contributions made by French social scientists such as Lenoir (1974) and Lefebvre (1974) by building on Marxist notions of socio-spatial exclusion as a necessary condition of capitalism. The term was first used by the European Commission in 1989 when the Council of Ministers requested a study of policies to combat social exclusion. In the Anglo-Saxon literature important contributions have been made by social geographers such as Sibley (1981, 1995). In a policy context, the term was appropriated by social democrats in the 1990s as a shorthand term for an inter-linked series of social problems. Following Hine and Mitchell (2000) and Burchardt et al. (1998), the following was adopted as a working definition (see also Table 1.1) for the Oxford University study:

> Social exclusion is a process which causes individuals or groups not to participate in the normal activities of the society in which they are residents.

Emphasis in the 'process' identified in the adopted definition is placed not only on income-based indicators of poverty but also access to health care, education, transport, labour markets, welfare provision, financial markets, housing and social and political networks. Particular emphasis is often placed on the processes and factors that limit participation in civil decision making (after Eisenstadt and Witcher, 1998).

A criticism of social exclusion is that it revives prejudicial notions of an underclass and the undeserving poor (Samers, 1998). In response to this, more recently, emphasis has been placed on the concept of social inclusion. The Centre for Economic and Social Inclusion (2002) highlights that this is "understood as a process away from exclusion, it is a process for dealing with social exclusion and integrating individuals into society". Broadening its definition the Centre states:

> Social inclusion is the process by which efforts are made to ensure that everyone, regardless of their experiences and circumstances, can achieve their potential in life. To achieve inclusion, income and employment are necessary but not sufficient. An inclusive society is also characterised by a striving for reduced inequality, a balance between individuals' rights and duties and increased social cohesion.

Luxton (2002) identified five critical dimensions for social inclusion:

- Valuing individuals or groups with sensitivity to cultural, ethnic, gender or age differences.
- Allowing and enabling individuals to make life choices and, if desired, to make a contribution.
- The right and support to be involved in decisions affecting oneself, family and community.
- Reduction of social distances and provision of opportunities, if desired.
- Resources to participate fully in community and society.

Transport contributes to all of these dimensions: the valuing of individual groups in a framework of sensitivity has to do with public exposure to the presence of such groups and public exposure requires travel and transport to accomplish visibility; the ability to make life choices only makes sense in the context where life choices can be accessed and this too necessarily involves transport and travel; the right to be involved in decision making requires presence at the decision table and this has clear physical and virtual access dimensions; the reduction of social distance involves interaction and interaction is accomplished by meeting in a range of social grounds all of which involve transport and travel; and finally, having access to the resources to the participate is predicated in previous interaction, negotiation and bargaining for resources, all of which are related to the extent to which mobility is either constrained or unconstrained (the immobile are in a very weak social bargaining position). The transport and travel dimensions of choice over lifestyle and determination of personal action space have been sadly neglected – most particularly by transport and traffic analysts. The reduction of transport planning into passenger car units lacks a sensitivity to address these types of issues: issues of culture and exclusion disappear in the orthodox model.

Within the social inclusion/exclusion discourse, which initially occurred outside of transport, there is an evident tension between the two terms, social inclusion and exclusion. The initial focus on exclusion – deficits – has moved towards a focus on inclusion – remedies – with many authors attending only to one side of the dynamic. In time, more rigour and more clarity may appear in the discourse: at present, the emphasis and focus at any one time has largely been led by the funding activity of major policy agencies. At present, policy agencies have begun to focus on the transport aspects of social inclusion/exclusion, however, the charting of the dynamics is far from complete. The particular focus of policy agencies has defined the scope and level of investigation of social inclusion/exclusion in transport policy without a high level academic appreciation of the total terrain.

Lucas et al. (2001) highlight four main ways that transport can contribute to social exclusion. These are: the negative impact of road traffic, inadequate public transport (see TraC, 2000), reduced or poor accessibility to basic facilities (for example, the work of Church et al. (2000) in London and Nutley and Thomas (1995) in rural areas) and travel poverty (Root, 1998).

Table 1.1 Definitions and Explanations of the Term 'Social Exclusion'

Definitions/Explanation	Source
Social exclusion is a process which causes individuals or groups, who are geographically resident in a society, not to participate in the normal activities of citizens in that society	Hine and Mitchell (2000)
Social exclusion is a shorthand term for what can happen when people or areas suffer from a combination of linked problems such as unemployment, poor skills, low incomes, poor housing, high crime, bad health and family breakdown	Social Exclusion Unit (2001a)
The condition of living in a society but not having the opportunity to participate in the normal activities of citizens in that society	Sinclair (2001)
People excluded from society are those who accept the goals of society in some loose sense but who either do not agree with the socially acceptable means to achieve these or find that the means are not available to them	Pearce (2001)
Social exclusion is… a multidimensional phenomenon… characterised conceptually as the process which prevents people from a full participation in the society	EU Business (2000)
Social exclusion is related to relative poverty where groups or individuals lack the resources to obtain type of diet, participate in the activities and have the living conditions and amenities which are customary, or at least widely acknowledged or approved, in the societies to which they belong	TraC (2000) after Townsend (1979)
Exclusion springs from the desire to belong while not being able to	McCluskey (1997)
A situation in which certain members of a society are, or become, separated from much that comprises the normal round of living and working in that society	Philo (2000)
The outcome of processes and/or factors which bar access to participation in civil society	Eisenstadt and Witcher (1998)

The early discussion of social inclusion/exclusion outside of the transport domain did have an ethnic dimension in the UK policy literature (Social Exclusion Unit, 2001a), however, even within the policy reports of the Social Exclusion Unit (2003) itself on transport the coverage of ethnicity is relatively thin. The 2001 Social Exclusion Report usefully highlights the fact that people from some minority ethnic communities are disproportionately exposed to risk of social exclusion. The document also reports that in the 1980s and the early 1990s, social exclusion intensified in this country in a variety of different ways. Furthermore, the report argues that high levels of exclusion also impose indirect social costs on the whole population. These costs include:

- reduced social cohesion as different areas, generations and minority ethnic communities are divided by radically different life chances,
- higher crime and fear of crime, for which social exclusion is a key driver,
- extra pressure on people working with excluded groups, and
- reduced mobility, as vulnerable people avoid certain parts of town or feel intimidated – rightly or wrongly (e.g. by groups of young people, beggars).

However, despite this recognition of the ethnic dimension to social inclusion/exclusion, the cultural specificities of ethnic travel behaviour and their interaction with and integration into mainstream public transport organization remains largely undiscussed. Two main exceptions can be found: the first is the work of the GLC Women on the Move initiative (1986) and the second and more recent source is the unpublished work of Beuret et al. (2000) undertaken for what is now the Department for Transport. Beuret et al.'s work is rich in understanding the interaction between transport, ethnicity and social inclusion/exclusion: their findings in many ways match and marry with the findings of our own research in Bristol and in Nottingham. Because their work is unpublished, it is useful to set out a catalogue of impacts and processes they have identified in this crucial social sector and this is provided in Chapter 3.

1.3   The transport policy environment: legacies and innovations

Researching transport and social exclusion has been rendered more difficult by the lack of adequate auditing of the traffic and transport environment in the past. Large changes have taken place in the public transport sector of the United Kingdom with very little mapping of the precise character of those changes: for example, there is no ready national data base which will enable statistics to be obtained on the loss of local services even in respect of low income communities where subsidized service provision is meant to be in place. With deregulation of the municipal public transport systems, and the contemporaneous rolling back of the public sector, there has been for many low income communities a major decline in the quality of public transport service provision. Concerns about the effects of poor public transport provision on residents of low income communities have been expressed on many fronts: this extract from a recent debate on bus company regulation in the House of Commons summarises some of the concerns:

Mr. George Mudie (Leeds, East): The Secretary of State is well respected, but he must conduct a review with seriousness; the matter is now past a joke. Whole communities in Leeds are being cut off, the elderly cannot go shopping, people cannot get to hospitals on public transport at weekends, and even the local hospice cannot be visited by public transport on weekends and evenings. That is clearly a ridiculous situation and public transport is becoming a joke. The bus companies are in it only for money, there is no public service and no public interest, and we are sitting on the sidelines looking like fools.

Mr. Darling: In Leeds, about which my hon. Friend knows a great deal from the time that he led the council there, bus patronage has increased. He is right to draw attention to the fact that some bus services late in the evening, in areas that are more difficult to reach, are not of the standard that they should be, but nor were they prior to deregulation in 1985. From the end of this year, we are extending the availability of the bus service grant to provide better, more bespoke services for areas without bus services, especially when there is thinner patronage than would justify a conventional route being run. Many innovative steps can be taken that will encourage bus use in areas that have not done well historically, either before or after deregulation. As I said previously, I am keeping the situation under review. I know that there are difficulties, which we want to sort out.
(House of Commons, Tuesday 21 October 2003,
http://www.publications.parliament.uk/pa/cm200203/cmhansrd/cm031021/d
ebtext/31021-01.htm#31021-01_spnew10)

It is unsurprising that most of the 'voices' in this volume wanted to discuss issues related to bus rather than rail travel, seeing bus as their only real mode of public transport, since, as the Transport 2000 Trust (2003: 21) reports:

The bus is the mainstay of public transport in the UK, accounting for nearly two-thirds of all public transport trips. In many parts of the UK, where rail plays a minor role in local public transport, then the bus **is** public transport.

For the few who talked about rail travel, it was seen as a challenging mode to use because of infrequent services, difficult interchange, poor station access, safety issues and deterrent fare structures:

We have two local stations, one that way a few hundred yards and one this way – Lawrence Hill and Stapleton Road. The trains don't stop in either of these stations on a Sunday. So you can't take the children out on a Sunday – it's ridiculous.
(Elderly white woman living in ethnically-mixed neighbourhood, Easton, Bristol)

You find that all the trains that come along from Gloucester, only three a day will stop here. Basically if you want to go like to the Severn Tunnel, all the trains that go past Stapleton Rd and Lawrence Hill will pass through but not

actually stop at these stations. If you are travelling to Easton and you want to be here before 9 o'clock in the morning, you've got the option of catching 0805 from Temple Meads which gets you here at 0815 at the latest or you come here gone 9 o'clock. It's an hourly interval to Severn Beach. The train system is not bad but it could be more accessible. If they stopped there once every half hour it would be better, it's not too bad in peak times now but really not good during the day.
(Asian man, 35 years old, Easton, Bristol)

You see there's a big shopping complex at Clifton that you could go to from here, get on a train, do your shopping and come back just more for a change of scenery and go somewhere else to shop. And also to go to Filton Abbey Wood but they don't stop often so it's not a local service. So the service is there but it is not adapted to the place it is running through.
(Elderly white woman living in ethnically-mixed neighbourhood, Easton, Bristol)

Basically the infrastructure is there but the service provision could be better.
(Asian man, 35 years old, Easton, Bristol)

On the buses, once you're 60 you can get a pass and travel for half fare but you can't do that on the trains. For example it is £6.60 to go to Weston on the train from here but if I go on the bus it is only £2.05 from town return.
(Elderly white woman living in ethnically-mixed neighbourhood, Easton, Bristol)

The local train service is a failure. Nobody uses it for a lot of reasons. Firstly they don't keep no times. Also there is a station on Stapleton Rd but how do people get up there? There's no lifts, nothing.
(Elderly Asian man, Easton, Bristol)

If I'm working in Somerset and want to get a train back I can only get it to Temple Meads station. I'd rather come to Montpelier (station) and walk down Stokes Croft. If you get to Temple Meads, you have to take a taxi because when I come back at 9:30-10:00 at night I don't really want to be walking through Bristol city centre to get a bus. I take a taxi, if they will take me, but now sometimes I have to rely on someone to pick me up at the train station or if I'm working with somebody they will give me a lift with their lift.
(Black Caribbean woman, 30 years old, St. Paul's, Bristol)

Even to get to the train station is a problem. You've got to get a bus into town and then a bus from town to the train station. So I have to ask a friend to take me.
(Black Caribbean woman, 20 years old, St. Paul's, Bristol)

We don't use train because they are expensive and there is no concession that we are aware of for older people.
(Elderly Asian man, Easton, Bristol)

On the Lawrence Hill (at the station) it is very dangerous to stand there, so scary, they mug you.
(Elderly Asian woman, Easton, Bristol)

And it is very expensive, double the bus and it takes you to Temple Meads which is far from the centre.
(Middle-aged Asian woman, Easton, Bristol)

At Stapleton Rd (station), they've spent some money on it. The problem is that they built the railway lines higher than the platform and people can't step up into the train.
(Elderly Asian man, Easton, Bristol)

In addition, public transport users frequently refer back to a better time in public transport:

It's not a very good bus service. It's where you want to go causes the problem. You've now got to go into the city to go out of the city whereas before you used to be able to go all over.
(Young white Mother, Clifton, Nottingham)

They're not so frequent as they used to be. Every time I've had to go to hospital you have to wait forever. Buses are often missing…
(Young white Mother, Clifton, Nottingham)

I have a disabled bus pass but if I go to the supermarket I have to rely on friends to take me there. There's no supermarket served by a bus from this area – there used to be one to ASDA but it's not around anymore.
(Elderly white woman living in ethnically-mixed neighbourhood, Lenton, Nottingham)

Once upon a time, many of us only took jobs if we could get there by public transport. Now a commuter is someone who comes into the city from a village by car. Perhaps they need to provide links that bring people in from the villages on public transport for shopping and business.
(Elderly white man living in ethnically-mixed neighbourhood, Lenton, Nottingham)

The bus station facilities now are not as good as the old ones – it needs a warm, staffed area, make it more like airports, it's as if the bus traveller gets short-changed all the time.
(Middle-aged white woman living in ethnically-mixed neighbourhood, Lenton, Nottingham)

(There are) certain areas that buses don't go after dark for fear of attack like Knowle and Southmead. When they had conductors on the buses at least the driver had some sort of back up. So money could be put into security measures on the buses. However, people know when there is a sign on a bus saying that it is covered by CCTV that it's not true.
(Elderly white woman living in ethnically-mixed neighbourhood, Easton, Bristol)

I'd get the bus into Bristol from my home if I could, but as bus services to the centre were removed from my village, I have to drive in. I think they should stop wasting money on strange traffic systems and subsidise more bus routes.
(Posted on BBC Bristol Message Board 14 Sep 2003 20:10 http://www.bbc.co.uk/cgi-perl/h2/h2.cgi?thread=%3C1061888923-23331. 6%40forum3.mh.bbc.co.uk%3E&find=%3C1061888923-23331.6%40forum 3.mh.bbc.co.uk%3E&board=talkbristol.bristolsoup1&sort=Te)

Some see little promise that public transport will be improved in the near future:

Large majorities believe the condition of the four main public services – education, the NHS, public transport and the police – has deteriorated, is deteriorating and, despite vast additional public expenditure, will continue to deteriorate.
(King (2003) commenting on the findings of a YouGov survey for The Daily Telegraph of 2,291 adults across Great Britain

The legacy of the rolling back of the public transport system for many low income housing estates and neighbourhoods is one of broken links on a number of fronts: health, wealth, social being and general quality of life. Severance issues which have limited impact where quality public transport links exist become overwhelming barriers to mobility and accessibility where walking or cycling are the options. For example, participants from the Inner City districts of St Paul's and Easton in Bristol highlighted problems of community severance (as a result of the M32), traffic congestion, parking difficulties, overcrowded public transport, with buses frequently full, and the poor standard of local facilities. They did not consider their neighbourhoods to be particularly accessible, even though deprivation indices data put them in the top quintile in terms of accessibility.[4]

Measurement of access to facilities and the difficulties of identifying levels of accessibility have been raised by Parkes et al. (2002) in light of individual neighbourhood dissatisfaction. Their analysis sought to determine what factors make people dissatisfied with their neighbourhoods and appeared to show that the role of access to facilities tended to be seen as of little importance. However, Parkes et al. state:

...the role which such facilities and amenities play indirectly – through, for example, enabling social interaction to take place – is little understood in contemporary society and is a weakness in our comprehension of urban neighbourhoods today.

When Parkes et al. (2002) go on to suggest that access questions are among the most problematic in the Survey of English Housing, asking people 'how well placed' their home is for certain facilities, they indicate that it is not only, as the road user charging/workplace parking levy study has revealed, the Indices of Deprivation that have shortcomings in their assessment of access, rather that this extends to other areas of social policy. Parkes et al. say that there is

> ...a lack of clear vision in the government's National Strategy for Neighbourhood Renewal (Social Exclusion Unit, 2001b) in relation to the type, quantity and quality of facilities and amenities which should be available at the local neighbourhood level in Britain.

In this light, there is an obvious need for professional measures of accessibility to more closely account for the real levels of access to key facilities being experienced by individuals from different social groups. The difficulties related to lack of appropriate indicators of accessibility have already been described. However, it is important to return to the issue of accessibility to discuss certain spatial factors that have been seen to affect access in the Oxford University research.

In examining neighbourhoods and social exclusion, Kintrea and Atkinson (2001) stated that one of the most obvious sources of neighbourhood effects (their definition: independent, separation effects on social and economic behaviour which arise from living in a particular neighbourhood):

> ...is the influence of the quality and availability of services which are consumed within local neighbourhoods.

The authors suggest further that:

> Studies of poor areas have consistently shown that services are indeed worse than elsewhere and their negative impact is intensified by limited mobility and poverty among residents.

With respect to the Oxford University study's findings, Kintrea and Atkinson's assertion that permeability of a deprived area must be increased if isolation is to be broken down is instructive. They argue that:

> Social isolation builds on physical isolation which is typically a product of a combination of factors; spatial distance from town centres, poor transport connections and stark boundaries such as main roads...

These sources of social isolation or, perhaps more appropriately, exclusionary factors have all been seen in this research: spatial distance from the centre was a factor for residents of Hartcliffe and Withywood in Bristol and Clifton in Nottingham, poor transport connections were described in Easton and St Paul's in Bristol and in The Meadows and Lenton in Nottingham. The boundaries between the communities of St. Paul's and Easton could not be more austere with a dual-carriageway running into a grade-separated junction over the M32 forming one of the main links between the two areas.

Reducing mobility and accessibility through the decline in public transport services represents an intensification of severance: physical, social and resource severance are all features of the low income environment. The issue now is to develop innovations in transport policy and practice which remedy this long developing, undesirable legacy. Within this frame, demand management policies can be usefully harnessed to improve the lot of low income and ethnic neighbourhoods. Parking control policies can be used to remedy the displacement parking which has accompanied the growth of car based commuting into Britain's congested city centres; road user charging and work place parking levies can be used to reduce the volumes of traffic passing through ethnic neighbourhoods adjacent to city centres; revenues from parking policies and traffic management measures can be used to improve the quality of poor public transport in low income neighbourhoods. New options in transport organization exist with the development of intelligent demand responsive transport which can be customized to meet cultural needs. However, all of these innovations carry the danger that they can also worsen the situation of ethnic residents and vulnerable groups. Parking controls set with inappropriate geographical boundaries can result in displacement parking into low income neighbourhoods. Road user charging can also result in displacement parking into low income neighbourhoods. Workplace parking levies have the same potential for resulting in displacement parking into low income neighbourhoods. All of these tools carry risks of worsening social exclusion if insufficient attention is paid to exemptions, concessions and auditing of the impacts. Demand responsive transport can be used as a transport tool for bypassing low income areas as well as a tool for servicing them, whereas fixed routes are open to all, demand responsive transport has a specified customer base.

1.4 Rethinking transport planning, restructuring transport policy: some suggestions

Auditing has become part of transport planning and practice over a range of operational areas: road safety audits, cycling audits; pedestrian and vulnerable road user audits, gender audits, accessibility audits, institutional audits (such as those undertaken by the World Bank in the transport sector – http://www.worldbank.org/transport/publicat/td-ut2.htm). That auditing is a useful and rapidly emerging tool within the transport domain is undoubtedly the case. Within this book and in relation to the development of demand management policy and its impact on ethnicity and social exclusion, we want to draw attention to the need for equity auditing in respect of new schemes (Grieco, 2002).

Our research indicates that there is a clear need for local authorities to consider the wider impacts of proposed charging schemes on different social groups. An equity audit would facilitate the examination of the key issues that need to be addressed in introducing such a scheme. By applying the audit checklist over various stages of the scheme, progress could be monitored and suitable interventions developed to enable improvements in equity in transport through the charging regime.

The parking displacement equity audit is a simple new auditing tool which is readily available to local authorities wishing to monitor and ameliorate displacement parking effects associated with the introduction of a new road user charging scheme: traffic wardens could be used to feed back real-time online information on parking violations resulting out of the introduction of the cordon. This is a potential instrument for a local authority to consider in the conducting of an equity audit (Rajé, 2002). Later chapters of this book discuss these new auditing opportunities more fully.

1.5 Conclusion: resolving transport exclusion, recipes for remedy

> Understanding self-perceptions and decision-making among ethnic minority groups is viewed as being particularly important in ensuring that future policies are effective and services culturally appropriate. (Proceedings from the Prime Minister's Strategy Unit, Strategic Thinkers' Seminar (May, 2003) @   http://www.number-10.gov.uk/su/social%20exclusion/downloads/ summary.pdf)

> Understanding how groups classified as 'socially excluded' perceive their own experiences and make decisions is a weak link in current thinking. (Proceedings from the Prime Minister's Strategy Unit, Strategic Thinkers' Seminar (May, 2003) @ http://www.number-10.gov.uk/su/social%20 exclusion/downloads/summary.pdf)

> Understanding race equality and diversity is as vital to the success of the Department as it is to all those who provide transport. Importantly, it enables us to develop policies and services that match peoples' needs and which makes a difference to peoples lives and experiences. It also helps us bring communities together, reinforce social values and understand better the links between transport and the opportunities to access services, work, culture and heritage for all members of our society. (DfT Race Equality Scheme 2003-2005 (2003) @ http://www.dft.gov.uk/   stellent/groups/dft_mobility/documents/page/dft_mobility _022396-01.hcsp#P28_549)

Despite the evident frustration of the low income public transport user about the circumstances and conditions of travel on socially necessary and even upon survival journeys and the net published rhetoric of public consultation on transport emanating from the offices of the most senior politicians, there is little evidence of on the ground change. Bringing an understanding of the workings of demand management into the public arena requires more local measures than simply web publishing the thoughts of strategic thinkers: at the very least, equivalent space web publishing the lived experience of public transport users as a consultation protocol would achieve a necessary balance.

The 'transport and social exclusion' analysis now developing within policy circles delivers the ground in which the movement has to be away from over simplistic 'travel awareness' policies towards more genuinely participatory planning. This need is most surely and strongly exhibited within the ethnic communities now resident and reliant on public transport in Britain. The quotations above taken from the web site of No 10 Downing Street indicate that thoughts have begun to turn in this direction. Remedies are on the horizon:

> Issues of race and ethnicity have come to the fore in British public life. The last few years have seen growing public condemnation of racially motivated violence and harassment, a hand-wringing debate on institutional racism following the publication in 1999 of the Macpherson Report into the death of Stephen Lawrence, proposals to combat racism and to ameliorate the lives of minority ethnic people, discussion of what nationhood and belonging means in a multiethnic society following the publication of the Parekh Report (Runnymede Trust, 2000), and a raw argument on the rights of asylum seekers feared to be swamping this crowded island. Most recently, after the street confrontations in 2001 in Oldham, Burnley, and Bradford, new issues have entered the national debate. These include alarm at the scale of ethnic deprivation and segregation in poor urban areas, growing Islamophobia and unashamed questioning of the cultural and national allegiances of British Muslims (reaching fever pitch after September 11), widespread moralising about what it takes to be British, and concern about the activities of racist organisations such as the British National Party (BNP), now increasingly tapping into anxieties of neglect and resentment among poor White communities.
> *(Amin, 2002: 959)*

The following chapters of this book focus primarily on the ethnic dimensions of the Oxford University study into transport and social exclusion in the context of new demand management policies. This is not because gender or lifecycle issues are deemed unimportant but because the ethnic dimension of travel and transport behaviour has been understudied, underrecorded, under-reported and, even where such studies exist, underpublished remaining in the policy drawers of Whitehall rather than being disseminated within the public domain.

Chapter 2 of this volume provides a condensed account of the development of the specific demand management tools intended for implementation in British cities with high levels of congestion and substantial ethnic presence.

We then turn to an extensive exploration of transport and social exclusion in relation to ethnicity in Chapter 3 making significant use of the 'voices' of our ethnic respondents in Bristol and in Nottingham – interestingly, and often overlooked, is the family and household connections between ethnic residents and what would normally be labeled 'white' residents of host community. In the opening voices of our book, the black single mother and white mother adjacent quotations were from a mother and daughter. The ethnic experiences of the daughter have consequences for the mother. It is a perspective rarely considered and one that we give some thought to and reflection upon in Chapter 3.

Chapter 4 provides a case study of the proposed road user charging scheme in Bristol and draws attention to issues of ethnic access in that context: it uses 'voices' to provide the baseline on current experience and to indicate the existing problems of public awareness and public acceptability. In Bristol, as we shall see, there was very low awareness of the proposed scheme: this raises issues of public consultation, consequent barriers to public acceptability and the need for national policy discussions on participatory planning.

Chapter 5 provides a case study of the proposed work place parking levy scheme in Nottingham, Here, as we shall see, there was a greater awareness of the proposed scheme which also involved building a tram line as part of the overall transport package accompanying the scheme: in this context the 'ethnic voices' still talked to existing public transport difficulties and concerns about the impact of the engineering works accompanying the building of the tram line generating nuisance in the streets of the ethnic neighbourhood. This chapter records the role that disruption and displacement parking play in reducing the initial acceptabilities of demand management measures from the perspective of the ethnic community. In Nottingham, this concern about disruption and displacement parking was expressed across the board of low income groups consulted.

Chapter 6 identifies the need for local audits before, during and after the undertaking of new schemes. In the case of road user charging, there are local dimensions of disruption and local aspects of impact and change in the situation of affected residents inside the cordon, around the cordon and beyond the immediate area of the cordon which require monitoring, remedying and compensatory revenue sharing in order to preserve or deliver social equity. Similarly, in the case of work place parking levy with accompanying new transport infrastructure measures, there are displacement effects which need monitoring. Furthermore, employees were concerned that lower level workers would bear the greater burden of any charges as employers would be more likely to provide higher level employees with packages that included waiver of the charges. Auditing displacement effects in both these contexts emerges as a key issue: this chapter offers a new tool for approaching such effects in the Parking Displacement Audit which can be used with compensatory revenue sharing to provide a level of resolution.

Chapter 7 concludes the volume with an overview of funding, financing and fine tuning measures which can be adopted to better meet the unwanted impacts of demand management in respect of social exclusion considerations.[5] The chapter argues that the use of revenues raised through demand management can be used to fund more participatory forms of planning which in their turn will generate greater public acceptance of these undoubtedly necessary measures. It finishes with a plea for ethnic inclusion in what have historically been highly exclusionary transport operation and planning practices.

---

1 renewal.net has been developed to fulfill one of the commitments of the Government's National Strategy Action Plan for neighbourhood renewal. Although the site has been developed by the Neighbourhood Renewal Unit, which is part of the Office of the Deputy Prime Minister, its aim is to provide an independent, evidence based view of what works and what doesn't in neighbourhood renewal. The documents commissioned for renewal.net have been produced by a team of independent experts. (Text taken from renewal.net)

2 This work came out of a scoping study carried out by Torben Holvad and John Preston, University of Oxford and Julian Hine, University of Ulster in 2000.

3 This was the partner project to a study carried out at Lancaster University 'Changing infrastructures, measuring socio-spatial inclusion/exclusion (CHIME)' which was conducted over the same period of time.

4 This is not surprising since measurement of accessibility for the Index of Deprivation 2000 used straight-line distances from the residence to the nearest facility (post office, food shop, GP and primary school only) and then aggregated these distances and ranked them according to magnitude. Also, the deprivation indices did not take account of other factors that affect an individual's levels of access such as quantity and quality of services, times, costs and car ownership.

5 For an insight into social exclusion considerations in relation to transport, see Transport Policy, 10, 4 – Special Issue on Transport and Social Exclusion

Chapter 2

Constraining Mobility, Demand Management and New Transport Policy Instruments: Road User Charging and Work Place Parking Levy

2.1 Introduction: the policy challenge – managing congestion whilst minimizing social exclusion

Charging schemes have the potential to make significant reductions in congestion and to improve the capacity, speed and reliability of public transport, but it is important that such schemes are designed to enhance the urban environment. Schemes which merely displace traffic from a city centre to suburban or inter-urban road networks may cure urban congestion at the price of urban decline, and will lead to problems elsewhere on the road network.
(Summary of Conclusions and Recommendations of Select Committee on Transport: First Report – Urban Charging Schemes (2003) @ http://www.parliament.the-stationery-office.co.uk/pa/cm200203/cmselect/ cmtran/390/39003.htm)

In-depth studies of public concerns about proposed road user charging schemes have revealed a number of issues, ranging from the reliability of the technology to a lack of acceptance of the principle of direct charging. However, the most pervasive and deep-seated concerns relate to the 'fairness' of the scheme. In political terms too, equity is a key issue, but one that has received relatively little attention by academics or practitioners.
(Addressing Equity Concerns in Relation to Road User Charging, Professor Peter Jones, Transport Studies Group, University of Westminster (UK) @ http://www.transport-pricing.net/jonel.doc)

The key argument of this book is that it is undoubtedly the case that traffic demand management tools are required to provide an efficient transport system within Britain. Equally important to the argument made here is the need to ensure that demand management tools are developed within a framework which pays full attention to issues of social equity. This chapter will provide an insight into the need for demand management tools and the background to the emergence of congestion charging as a policy instrument in the UK. The key aim of the Oxford University research on which this book itself is based was to examine the impacts of road user charging/workplace parking levy on social inclusion and exclusion. Individuals from ethnic minorities, amongst other social groups such as women, the young and elderly, may be particularly affected by congestion charging due to a number of factors such as low income, working patterns, current high rates of social exclusion, high reliance on car-based travel for some journeys and the non-feasibility of public transport for some journeys. The research sought to explore the range of impacts of different charging schemes and scenarios on people's travel behaviour, time organisation and socio-economic activity. It did so by interviewing and conducting focus groups and collecting travel diaries with members of the affected public in two cities in Britain, Bristol and Nottingham, where demand management schemes were under consideration or in development. Details of the research process and methodology are provided in Appendix 1 and, in Chapters 4 and 5, the proposed operation of these schemes is considered in detail. In this chapter, we have a glimpse of the views of Bristol and Nottingham residents into the workings of Road User Charging and Work Place Parking Levy. Within this book, we are primarily interested in the ethnic dimension of perceptions and experiences of transport and its management and the focus will be placed primarily upon the displacement impacts of road user charging and work place parking levy on ethnic neighbourhoods adjacent to proposed demand management schemes.

To begin with the wider context, the technical definition of transportation demand management is:

> Transportation Demand Management (TDM): Strategies that result in more efficient use of transportation system capacity. Road pricing is a major category of TDM.
> *(TDM Encyclopedia Victoria Transport Policy Institute @*
> *http://www.vtpi.org/tdm/tdm61.htm)*

> Transportation Demand Management: A variety of strategies…. to influence travel behavior by mode, cost, time, or route in order to reduce the number of vehicles and to provide mobility options. TDM strategies are often applied to achieve public goals such as reduced traffic congestion, improved air quality, and decreased reliance on energy consumption. TDM strategies are also used by employers to reduce overhead costs, enhance productivity, and address other business problems such as employee turnover.
> *(National Center for Transit Research (NCTR) located at the University of South Florida @ http://www.nctr.usf.edu/clearinghouse/tdmterms.htm)*

Discussions of demand management, primarily focused upon policy induced reduction in car based mobility, have developed in the United Kingdom and in Europe against increasing levels of congestion and in the context of an increasing awareness of the environmental damage produced by high levels of car based mobility. MORI (2001) states that the single most important transport issue for the public reported across the country is congestion. Research carried out on behalf of the DfT concurs with this view:

> Congestion is an important issue for most drivers. It tends to be among the top two or three concerns among the problems seen to face motorists, and in some areas it emerged as the most important.
> *(Hedges, 2001)*

There is other evidence to support this finding. For example, Transport and Travel Research (2002) found that reducing congestion was a highly rated matter in terms of perceived importance amongst respondents from the general public to a survey carried out in Nottingham – one of our field study cities. MORI's survey findings suggest further that people are prepared for both radical action and higher public investment to alleviate congestion problems.

Given this background, the need for policy tools to tackle congestion is clear. As part of the Transport Act 2000, the Government introduced legislation to enable traffic authorities to introduce Road User Charging in all or part of their area, or on particular roads. The legislation allows authorities to introduce a compulsory charge for driving in a designated zone or on specific roads, either all the time or at specified times. With respect to Workplace Parking Levy, the government introduced legislation that allows traffic authorities to charge on private, non-residential parking at the workplace across all or in designated parts of their area. Schemes introduced should be designed to reduce congestion or traffic growth, or to meet other objectives contained in a local transport plan.

The essence of demand management transport policy in the United Kingdom is the constraint of car-based mobility. The case for demand management is not essentially a social exclusion rationale. But demand management has many social exclusion aspects and consequences which require attention. Precisely whose mobility is constrained by demand management becomes a public acceptability issue: adverse impacts upon vulnerable groups may not only impose difficulties for those groups but increase the costs to the various social and welfare systems concerned with making provision for the vulnerable. Consequently the issue of exemptions and concessions very rapidly becomes of concern in the implementation of any particular scheme – a factor that was well demonstrated in the recent introduction of congestion charging in the city of London.

In addition to the social allocation of concessions and exemptions, the issue of the use of revenue generated by congestion charging has received substantial comment. The MORI (2001) report states that a majority of people are in favour of both congestion charging and motorway tolling when revenue is 'ploughed back into transport'. With reference to social exclusion, the survey found transport is perceived as critical to life as it is seen as the key to freedom, independence, access to work and social opportunity. In this context, policies which constrain mobility must be approached with due caution: both direct and indirect effects of such policies require careful consideration.

In order to have the opportunity to participate in society, individuals must be able to access key services with relative ease. The decline of local service provision associated with the movement of facilities such as grocery shops to out of town locations (as a result of land-use planning policies in the 1980s and early 1990s), necessitating increased travel, has resulted in certain groups finding participation in the normal activities of their society difficult. Such effects are exacerbated by the fact that, in addition, nearly one in three households do not have access to a car (Social Exclusion Unit, 2003) and viable public transport alternatives may not be available. As we have already seen, these effects are increasingly discussed in the language of social exclusion.

The Social Exclusion Unit (2003) states that:

> Some people, in both urban and rural areas, cannot reliably get to key places in a reasonable time.

Accessibility or 'reachability' of services and key locations is one aspect of social exclusion: exposure to injury or personal insecurity and crime is another. Analysis of Department of Transport, Local Government and the Region's data (now the Department for Transport) (Graham et al., 2002) suggests that the most deprived local authority districts have up to five times as many child pedestrian accidents per child as the least deprived.

In addition, turning specifically to the issue of ethnicity, a recent report (DfT, 2001) indicates that Asian children in the UK are involved in up to twice as many pedestrian accidents as the national average and that the risk is greatest in families where parents or carers are unfamiliar with UK traffic conditions, particularly for those Asian families who are more likely to be recent arrivals from abroad. This study only looked at Asian children but suggests that children of other ethnic minority backgrounds may be similarly compromised. Clearly then for ethnic minority families who are often concentrated in inner city suburbs where traffic flows are heavy and consequent exposure to risk great, the car can become a major cause of social exclusion as a consequence of the greater ethnic exposure to the risk of road traffic accidents.

Transport is capable of creating radical social change (Root, 1999) – in the case of the coupling of rising car ownership with attendant policy permitting the flight of local services away from local neighbourhoods, this change has not always been positive for some, more vulnerable, social groups. This book seeks to explore the extent to which transport generally, and congestion charging in particular, can create positive social change at a local and individual level. In so doing, however, it also investigates the degree to which transport (and congestion charging) can be exclusionary for some groups of people if appropriate remedial policy measures are not researched and adopted.

Summarising the state of the field, congestion and environmental evaluation have led to the development of policies which constrain car based mobility within the UK and Europe – demand management – and as these policies grow in importance and extend the areas that they affect caution must be taken to ensure that such measures do not have unintended exclusionary effects. To this end, tools for monitoring, auditing and remedying adverse effects must be an integral part of the demand management package if it is to find public acceptability. Not least of those impacts requiring monitoring is the impact upon ethnic communities.

2.2   Road user charging and work place parking levy: a new suite of policy tools for managing congestion

Transportation demand management measures are wider than road user charging and work place parking levy: in the United States, for example road space has been reorganized to dissuade single occupancy of vehicles as a transportation demand management tool and, in Washington State, commuter trip reduction plans are a normal policy tool with detailed specification of measures to be taken by employers to ensure no reduction in workforce access to the workplace. Employers are expected to be active in promoting car pools, alternative journey methods to single occupancy car journeys. The discussion of transportation demand management is, thus, a major discussion in the United States albeit in a different form from the European approach. A useful summary of an American demand management package of policies and tools can be viewed on line for Washington State @ http://www.wsdot.wa.gov/tdm/tripreduction/CTRLaw.cfm): the policy philosophy behind the adoption of demand management is explicit and openly provided to the residents of the state through web based information services:

> The legislature finds that automotive traffic in Washington's metropolitan areas is the major source of emissions of air contaminants. This air pollution causes significant harm to public health, causes damage to trees, plants, structures, and materials and degrades the quality of the environment. …Increasing automotive traffic is also aggravating traffic congestion in Washington's metropolitan areas. This traffic congestion imposes significant costs on Washington's businesses, governmental agencies, and individuals in terms of lost working hours and delays in the delivery of goods and services. Traffic congestion worsens automobile-related air pollution, increases the consumption of fuel, and degrades the habitability of many of Washington's cities and suburban areas. The capital and environmental costs of fully accommodating the existing and projected automobile traffic on roads and highways are prohibitive. Decreasing the demand for vehicle trips is significantly less costly and at least as effective in reducing traffic congestion

and its impacts as constructing new transportation facilities such as roads and
bridges, to accommodate increased traffic volumes.

In the United Kingdom, demand management has been a narrower policy
field than in the United States and until very recently was not an area in
which any major legislation had taken place. High occupancy vehicle lanes
are backed by law in the United States whereas high occupancy vehicle
lanes are only just beginning to feature in the UK transport environment.
Green and workplace travel plans – the equivalent of commuter trip
reduction plans – have recently become part of the British transport
planning environment but are weakly enforced and with a very low public
profile. The British literature has primarily focused on road user charging
and this literature is thicker than the literature on work place parking levy.
Indeed, road user charging has been accompanied by major discussions of
political acceptability for over a decade in Europe (Grieco and Jones,
1994), however, workplace travel plans have a low profile and work place
parking levy has received little formal policy discussion accompanied by
largely disgruntled journalistic coverage.

Ison and Rye (2002) provide a useful comparison of work place travel
plans and road user charging as demand management measures in the UK
context:

> The nature of travel plans and road user charging are fundamentally
> different. Road user charging invariably provokes major hostility, tends to be
> higher profile and requires introduction on an area wide basis. Travel plans
> on the other hand are generally less contentious and are introduced on an
> individual employer basis.
> *(@ http://www.imprint-eu.org/public/Papers/Imprint3_Ison_Rye.pdf)*

In Britain, whereas previously road user charging had been part of the
professional transport discourse the advent of a New Labour government
put road user charging on the political transportation agenda:

> In July 1998 the United Kingdom Government issued its White Paper "a
> New Deal for Transport – better for everyone", the first comprehensive
> transport policy for nearly thirty years. It sets out policies to integrate not just
> the different transport modes but also transport with Education, Health, the
> Environment, Business and Industry. A key policy was the deployment of
> road user charging to attack congestion and pollution.
> *(Sampson for the DETR, 2000, @*
> *http://www.torino2000.itscongress.org/call/acrobat/sample.pdf)*

Despite the claim 'better for everyone' this landmark document does not
provide an adequate discussion of equity and vulnerable groups in relation
to the new demand management techniques and tools it introduces. A
useful and readable source of information on the new suite of policy tools is

to be found in the information packages of the Commission for Integrated Transport. The Commission for Integrated Transport (CfIT, 2002) provides a summary of the introduction of Road User Charging in the UK transport policy environment:

> First described in the DfT publication "Breaking the Logjam", the government introduced legislation as part of the Transport Act 2000 to enable traffic authorities, including the Mayor for London, to introduce road user charging in all or part of their area, or on particular roads. The legislation allows authorities to introduce a compulsory charge for driving in a designated zone or on specific roads, either all the time or at specified times. It is designed to cover both cordon charging and area licensing. For up to 10 years from implementation revenue from the schemes can be retained by the local authorities to fund wider improvements to the transport network in the area.

The first Road User Charging scheme in the UK was launched in the historic district of Durham in October 2002 and was confined to one street: the scheme has reduced traffic levels by over 80%. Interestingly, at the heart of this scheme was the upgrading of a public transport link (which is poorly used). The effectiveness of this quality public transport link combined with a congestion charge for car use in reducing traffic levels also had the consequence of precluding the scheme being a major revenue earner. Cities such as Edinburgh, Leeds and Bristol are considering introducing the policy and London's scheme started on 17 February 2003. London's adoption of road user charging gives us a view of what future legislative instruments will be used in British demand management schemes starting from the Transport Act 2000 through recent amendments on penalty charges: these are available on line (for example, Penalty Charge Amendment @ http://www.hmso.gov.uk/si/si2003/20030109.htm).

With respect to Workplace Parking Levy, the Commission for Integrated Transport's recent fact sheet on congestion charging (CfIT, 2002) reports:

> The government has introduced legislation that allows traffic authorities (including the Mayor for London) to introduce a charge on private, nonresidential parking at the workplace across all or in designated parts of their area. Schemes should be designed to reduce congestion or traffic growth, or to meet other objectives contained in a local transport plan. As with road user charging (above), revenues can be retained by the authority for up to 10 years from implementation of the scheme and spent on wider transport improvements in the area.

Local authorities in cities such as Chester, Reading and Milton Keynes have been considering the possibility of introducing Workplace Parking Levy. The most advanced planning for such a scheme is in Nottingham where implementation is projected to take place by around 2005 (CfIT, 2002).

2.3 Realising the limitations: equity issues and operational solutions

Though very little appears to have been written about road user charging and social inclusion, this section of the book seeks to give an overview of road user charging and current thinking on this policy in order to help ascertain its likely contribution to social inclusion/exclusion. Much of the literature focuses on economic and scheme design issues of RUC. Some of the articles provide a synopsis of the international position with regard to the status of various schemes in countries such as Norway, Singapore and the United States.

In relation to the impacts of road user charging, the emphasis in the papers and other materials reviewed (see Table 2.1) has been on describing uses of revenue in applications that imply a concern about the need for mitigation measures. Sinclair (2001), reviewing the proposed Road User Charging Scheme in Edinburgh, states it is important to establish the effect of road user charging on the behaviour and household budget of the socially excluded and people vulnerable to exclusion. The author suggests that to benefit lower income groups, road user charging revenue could be invested in public transport to make general improvements, increase/ safeguard its affordability and maintain/enhance its accessibility. In her later report, Sinclair (2002) states that, since the proposed charging rate is the same for all income groups in the case she studied, some individuals with low incomes (for whom public transport is currently not a realistic option) will be hit harder by introduction of congestion charging. As with much of the literature, ethnicity is not a focus within this research. Amongst those likely to be affected are shift workers and those people who have to undertake multi-purpose trips with time constraints eg. single parents. The use of revenue from road user charging is also discussed in the ECOPLAN report (1997) which suggests that, among other measures, this financial resource could be used for specific projects concerned with public transport and walking/cycling.

Table 2.1 Main Materials Reviewed on Road User Charging

No.	Article
1	Ahlstrand, I (2001)
	The Politics and Economics of Transport Investment and Pricing in Stockholm. Journal of Transport Economics and Policy, 35, 3, September 2001, pp. 473-489
2	Boot, J, Boot, P and Verhoef, E (1999)
	The Long Road Towards the Implementation of Road Pricing: the Dutch Experience. Paper presented at ECMT/OECD workshop on Managing Car Use for Sustainable Urban Travel, 1-2 December 1999
3	Bristol City Council (1999)
	EURoPrice Technical Paper 1: Priority Policy Issues Report
4	City of Edinburgh Council (2000)
	EURoPrice Technical Paper 3: Priority Policy Issues Report
5	ECMT (2001)
	Implementing Sustainable Urban Transport Policies. 30 April 2001, CEMT/CM (2001) 13
6	ECOPLAN (1997)
	Combined Road Pricing/Car Park Charging System for the city of Bern
7	Eliasson, J and Mattsson, L (2001)
	Transport and Location Effects of Road Pricing: A Simulation Approach. Journal of Transport Economics and Policy, 35, 3, September 2001, pp. 417-456
8	Enoch, M (2001)
	Arriving at a Transport Utopia by an Alternative Policy Route. Paper prepared for the IRNES Workshop, Machester Metropolitan University, November 2001
9	European Commission (2002)
	Urban Transport Pricing in Europe. Information taken from National Reports at http://www.transport-pricing.net/nationalreport.html
10	Goh, M (2002)
	Congestion management and electronic road pricing in Singapore. Journal of Transport Geography, 10, 1, March 2002, pp. 29-38
11	Goodwin, P (1998)
	Unintended Effects of Transport Policies in Transport Policy and the Environment. Banister, D ed. Son, London, 1998, pp. 114-130
12	Goodwin, P (1989)
	The 'Rule of Three': A possible solution to the political problem of competing objectives for road pricing.' Traffic Engineering and Control, October 1989, pp. 495-497

13   Guller, P (1999)

Pricing Measures Acceptance. Preliminary analytical results of the Prima-Project of DG VII/EU. Paper presented to ECMT/OECD Workshop on Managing Car Use for Sustainable Urban Travel, 1-2 December 1999, Dublin

14   Leach, G (2001)

More Roads and Road Pricing – The Way to go? Institute of Directors Policy Paper

15   May, A (1999)

Making the Links; Car Use and Traffic Management Measures in the Policy Package. Paper presented to ECMT/OECD Workshop on Managing Car Use for Sustainable Urban Travel, 1-2 December 1999, Dublin

16   Schlag, B (1997)

Public Acceptability of Transport Pricing. Dresden University of Technology, Germany

17   Sinclair, F (2002)

Assessment of the Effects of Congestion Charging on Low Income Households in Edinburgh – An analysis of Scottish Household Survey Data. PROGRESS. TRI, Napier University

18   Sinclair, F (2001)

Assessment of the Effects of Road User Charging and the Transport Investment Package Proposals on Social Inclusion – Recommendations for Consultation and Appraisal. PROGRESS. TRI, Napier University.

19   Small, K (1992)

Using the Revenues from Congestion Pricing. Transportation, 19, 4, 1992, pp. 359-381

20   STA (City of Rome's Mobility Agency) (2000)

EURoPrice Technical Paper 2: Priority Policy Issues Report

21   TDM Encyclopedia (2002)

Road Pricing: Congestion Pricing, Value Pricing, Toll Roads and HOT Lanes. Victoria Transport Policy Institute

22   Transport and Travel Research (2000)

Urban Road User Charging Scheme Design Principles and Policies – Report on Literature Review. Prepared for DETR.

The importance of social and political acceptance of road user charging is widely appreciated in Europe and is underlined by Rome's Transport Authority, STA (2000) which asserts that the viability of road pricing depends upon perceived benefits and justification given for the introduction of the scheme in a particular area. In addition, STA points out that it is necessary to consider whether certain groups will be marginalized through the introduction of charges for road use.

The effects of workplace parking levy on social exclusion are more complex to analyse compared to those associated with road user charging. Charges are incurred by employers, so in the first instance would not have impacts on the access to facilities and employment for individuals. Impacts on social inclusion/exclusion will therefore be determined by how employers will respond to the workplace parking levy charges. The assessment of available literature on workplace parking levy seeks to develop a clearer understanding of the policy and its potential effects.

While much has been written about road user charging in terms of scheme design or general impacts, relatively little can be found on workplace parking levy. The available literature suggests that employers play an important role in determining employees' travel and, in general, workplace parking levy is viewed as being easier to implement than road user charging. However, May (1999) states that it is difficult to judge how effective workplace parking levy charges will be since some employers will not pass the charge on to drivers or reduce their parking stock. Research by Ison and Wall (2002) looks at how acceptable and effective workplace parking levy is likely to be compared with other policy options and indicates that ring-fencing and clear allocation of revenue raised are important. Their findings also imply a pervasive perception that workplace parking levy is not as effective as introducing road user charging in urban areas. Craig (2000) reports that investigation of a charging structure showed that workplace parking levy could have an impact on traffic levels if set at the right level, although there was concern that the policy would not appear to encourage modal shift if employees tended to transfer to parking on surrounding streets where possible. Nevertheless, modelling carried out in California (Metropolitan Transportation Commission, 1999) indicates that a workplace parking charge would appreciably reduce vehicle hours of delay by encouraging people to car pool or switch to other forms of transport.

Gerrard et al. (2001) state that there is considerable interest in how the economic viability of urban areas will be influenced by policies such as workplace parking levy but recognize that little is known about which businesses are most likely to be affected, what social benefits business leaders perceive will accrue from such schemes and how businesses are likely to respond to the introduction of such a policy. Their paper seeks to address these issues and indicates that the majority of businesses expect workplace parking levies to lead to a reduction of both traffic pollution and traffic congestion. In addition, Gerrard et al. report that businesses expect workplace parking levies to reduce their own profitability and affect the economic prosperity of a city negatively.

The details of scheme design will be crucial in determining the extent to which road user charging or workplace parking levy contribute to social exclusion/inclusion processes. Several authors have highlighted the importance of how revenues from road pricing are used in order to ensure public acceptance of such schemes. In particular, Jones (1998) notes on the basis of public attitude surveys that

> ...road pricing will not be publicly acceptable unless the money raised is hypothecated for local transport and environmental projects.

In the context of social exclusion impacts, allocation of revenue from road user charging/workplace parking levy will be a key issue in preventing adverse impacts. The issue of hypothecation was emphasized by the Government in Breaking the Logjam (DETR, 1998) and was supported by a majority of the responses received in the consultation exercise (DETR, 2000). It is now enshrined in Schedule 12 of the Transport Act 2000. There could be considerable scope to reduce social exclusion and promote social inclusion by using the revenue collected from the road user charging/workplace parking levy charges to address the accessibility needs of persons experiencing a transport disadvantage in order to carry out their preferred or actual activities. In the context of gender, ethnicity and lifecycle a whole range of policies may be required to accommodate for the variation in problems across individuals. Indeed, the solution to accessibility problems may not even be transport-related, as problems can exist in the form of constraints at either of the journey ends, e.g. insufficient childcare facilities.

Furthermore, transport needs for the socially excluded could be addressed through initiatives aiming at reducing the need for travel, thereby achieving two key policy objectives; sustainability and social inclusion. The TDM Encyclopedia states that road pricing has two general objectives: revenue generation and congestion management. The Encyclopedia suggests that, although tolls represent a greater financial burden on lower-income than on higher-income motorists, they are not necessarily regressive since this attribute would depend on how much lower-income drivers use roads that have tolls, the quality of travel alternatives and how revenues are used. The authors recognize that lower-income motorists are sometimes willing to pay for time savings, indicating that pricing strategies that prioritize trips can provide a transportation choice that is valued by drivers of all income levels. In this way, congestion charging can improve basic mobility by giving priority to high value trips.

Enoch (2001) states that pricing is a type of mechanism that aims to regulate traffic to 'economically efficient' levels through reduction in vehicle use. However, he perceives that one problem with pricing policies is that equity is an issue, with those who can afford to travel benefiting at the expense of the less well off. In terms of parking, Enoch suggests that pricing, as opposed to regulating, is better able to match demand with supply, contributes to reducing congestion and provides a source of revenue. Nevertheless, it is interesting to note that Day (2002) reports that business directors in the UK are sceptical about the Government's Ten Year Plan for transport being able to reduce road congestion, with over 80% feeling it was unlikely or very unlikely to do so.

Demand management policies have clearly raised equity issues within the discussions of the transport planning community and these discussions have provided a menu of resolutions: hypothecation, compensatory public transport improvements, exemptions and concessions. Within this book, new items are added to this already growing list. Equity audits, displacement parking audits, compensatory revenue sharing for adversely impacted geographical areas to fund compensatory public transport provision such as demand responsive transport schemes, new levels of consultation and above all adequate consultation with ethnic communities in the context of recognizing and recording cultural differences in travel behaviour and travel patterns.

2.4   Working the boundaries: controlling the cordon

The demand management schemes we examine in this book have ethnic neighbourhoods adjacent to their boundaries: they will be the recipients of negative displacement effects from the proposed and under construction demand management schemes. This simple geography has inequity dimensions. Metz has characterized transport in Britain at present as 'a relatively egalitarian domain' (Metz, 2002). Our work, and that of others, disputes this and highlights the social inequalities that exist in the transport sector – most particularly in relation to ethnic and other vulnerable groups. The study reported on in this book found that poor access to transport is substantially restricting mobility amongst ethnic and other groups. People's description of public transport as poor in Bristol was shared with Nottingham though not as strongly. In both cities, it was clear that the base upon which demand management would build had substantial transport inequities exacerbated by unreliable public transport. It has been argued in other studies, most notably that of Lucas et al. (2001), that traffic restraint, particularly if based on charging, will exacerbate social exclusion. Taking these factors and views on board, it is important that the precise location of road user charging cordons are evaluated in terms of their impacts on inequity and that the boundaries of workplace parking levy schemes receive similar auditing.

Although we agree that the so-called deprived sections of the population are more dependent on car travel than is popularly thought, our research also suggests that amongst the groups that we surveyed only around one in ten people would be affected on a particular day by the proposed charging regimes. Hence for the people we surveyed, use of the revenue raised was more of an issue than the manner in which the revenue was raised, although there was some discussion of exemptions for, for example, low income workers and community car clubs. Rather than concentrating on disbenefits of charging policies, participants were more inclined to highlight the inadequacies of bus services and the limited scope of demand responsive services, as well as the poor street environment for pedestrians and cyclists. There was a realisation amongst many that revenue hypothecated from transport charging could help alleviate these problems, although there was also some scepticism about the ability of Local Authorities to deliver the desired policies.

In summary then, the baseline evidence provided by focus group members in Bristol and Nottingham – presented as 'voices' in later chapters of this book – does not reflect the equity Metz perceives. Indeed, in contrast to his view that 'Introducing congestion charging on a scale sufficient to improve efficiency substantially would result in a big move away from equity,' the empirical findings suggest that effective investment of revenues from charging can do much to restore equity by transferring money from the car driving sector to the more vulnerable groups who rely heavily on other modes. In this way, investment of RUC and WPPL revenue can facilitate assured access to services and encourage participation. The people taking part in this research have made it clear, however, that improvement in their experience of transport is largely needed on the lower end of the investment scale – they do not want to see large scale engineering projects but need adjustments and enhancements to existing services: more flexibility in demand responsive transport and extension of its hours of service, accessible bus fleets, readily available and appropriate forms of public transport information, security improvements.

In order to ensure that the needs identified by the groups of people involved in this study are explored and amelioration of their difficulties prioritised, a tool such as an equity audit should be considered for all Road User Charging and Workplace Parking Levy schemes. This requires repeated updating in order to enable the adjustment and fine tuning of schemes in terms of necessary exemptions, concessions and compensatory revenue sharing for the improvement of public transport. In this way, not only will baseline issues be obtained but progress towards equity will be measurable with any adjustments needed to suppress rising inequities being captured expeditiously.

As we shall see, the findings of this research indicate that Road User Charging and Workplace Parking Levy will have varied levels of impacts on the different groups studied. Nevertheless, the importance of the potential for hypothecated revenue to bring positive change, even at the lower end of the investment scale, to the people involved in this research cannot be overstated. The concept of social exclusion assists in determining the impacts of demand management in its recognition that transport is a social as well as a spatial process (Root, 2003). Public acceptability of demand management measures is more likely where equity issues have been explicitly considered within schemes and attended to. The recycling of revenue can promote social inclusion: demand management revenues provide a new source of finance for change in the public transport sector. Given these considerations the lack of incorporation of social inclusion as a criterion in the New Approach to Appraisal and related methodologies with Britain's transport policy legislation is disturbing as this means that there has been a lack of empirical studies on the social inclusion impacts of transport polices. It is hoped that our work constitutes a small step towards filling that gap, although clearly more work is needed. At the very least, one consequence of this failure is a lack of attention to the spatial patterning of urban ethnicity and the failure to achieving a policy recognition of boundary effects.

2.5 Conclusion: expanding the public discourse on demand management measures

The quotation from Professor Peter Jones that opened this chapter raised the important issue of fairness in the development of demand management transport regimes. Fairness is not accomplished simply by logically deducing impacts and effects: fairness is dependent upon public engagement and discourse about the character of the civil agreement which guides transport organization. Failure to engage with all the relevant stakeholders no matter how difficult this engagement is contains within it the seeds of failure. In the United States, travel plans failed in all but the state of Washington because the necessary discourse with employers did not precede legislation. In the United Kingdom, road user charging, work place parking levy and travel plans will all meet substantial obstacles if public consultation and the subsequent adjustment of plans in response does not take place.

The difficulties that such policies face in the United Kingdom are going to be strongly determined by the quality of public transport and the public's perception of public transport quality. MORI (Page, 2003) recently found that of all major public services 'public transport is that on which the public is consistently most negative – with more expecting deterioration than on any other major service area':

The public are more negative about transport than other public services..

Q   Thinking about... over the next few years, do you expect it to get better/worse?

Net
proportion
who expect it
to get
better

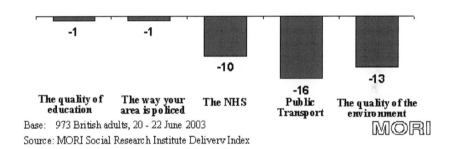

-1	-1			
		-10		
			-16	-13
The quality of education	The way your area is policed	The NHS	Public Transport	The quality of the environment

Base: 973 British adults, 20 - 22 June 2003

Source: MORI Social Research Institute Delivery Index

MORI

(MORI @ http://www.mori.com/pubinfo/bp_road-to-nowhere.shtml)

The two locations in which road user charging has now been adopted, Durham and London, have both paid attention to the quality of public transport in the implementation of their schemes. Given the MORI statistics, other authorities wishing to implement demand management policies will be obliged to do the same.

Chapter 3

Ethnicity and Transport:
A Neglected Dimension

3.1 Introduction: ethnicity, an unexplored social transport dimension

Most of our wives rely on other people all the time. I didn't have a car before and I had to ask someone to make a shopping for me. Now I bought a car but if I could have good transportation otherwise I would not buy a car.
(Somali Muslim man, 29 years old, Easton, Bristol accompanied by wife in burkah wearing the veil)

67% of all people from minority ethnic groups live in the 88 most deprived local authority districts compared with 40% of the general population.
(DfT Race Equality Scheme 2003-2005, 2003 @ http://www.dft.gov.uk/ stellent/groups/dft_mobility/documents/page/dft_mobility_022396- 01.hcsp#P28_549)

As well as significant areas of deprivation which are often poorly served by the transport networks, specific groups amongst Londoners are excluded from fully benefiting from transport services. These can include, for different reasons, women, young people, older people, people on low incomes, black and ethnic minority people, and most acutely of all, disabled people.
(Draft Transport Strategy for London, 2003 @ http://www.london.gov.uk/ approot/mayor/strategies/transport/pdf/2challenges.pdf)

In mainstream British transport analysis, the assumption is that all citizens have equal and ready access to public movement and public space. The advent of a multi-cultural society has not been accompanied by any refinement of the analysis of individual entitlement to movement. Yet within the United Kingdom for example, there are minorities present who practice the social seclusion of women. The number of women secluded in Britain has neither been investigated nor is it known. Whilst the research into transport and social exclusion in Bristol and in Nottingham did not provide examples of the absolute seclusion of women, references are made by respondents to the partial seclusion of Asian women. A Muslim woman in Bristol reports on her preference for using Muslim driven taxis whilst an Asian shopkeeper in Bristol reports on the chaperon practices surrounding his daughter's mobility even as a professional woman. A Somali man reports on shopping being undertaken for his family not by his wife, as we would find in mainstream British society, but by himself or others performing the service for him. Significantly, in the research Asian women showed a propensity to be car passengers. Examining Asian women's travel behaviour as car passengers has an added importance if viewed from the perspective of cultural practices of seclusion within the Asian Islamic community. The fear of assault in the transport environment for Asian women is not simply that of physical assault but of the culturally feared and proscribed assault on honour.

This chapter will highlight the distinct lack of relevant data about ethnic minorities and transport both in the UK and abroad. The official UK statistics website 'National Statistics' (www.statistics.gov.uk) provides Census 2001 data which indicate that 9% of the population of England were from ethnic minority groups. Yet, there is still very little publicly-available research on how ethnicity influences transport choice and a paucity of data on the patterns of travel of the various ethnic minority communities in the UK. This research has found voiced evidence that indicates that ethnicity is an important factor in the experience of transport.

Ethnic minority populations are not evenly distributed around Great Britain, but tend to be highly concentrated in the more urbanized parts of the country. According to the then Cabinet Office's Social Exclusion Unit (now part of the Office of the Deputy Prime Minister) (2000):

> In comparison to their representation in the population, people from minority ethnic communities are more likely than others to live in deprived areas; be poor; be unemployed, compared with white people with similar qualifications; suffer ill-health and live in overcrowded and unpopular housing.

Indeed, in terms of employment, The Prince's Trust (2001) has suggested that belonging to a minority ethnic group makes an individual 2.2 times more likely to be unemployed than a white person. The differences in economic activity rates vary amongst males and females and amongst different ethnic groups. The Trust suggests that there is a multitude of explanations to account for the difference in economic activity rates, for example, for South Asian women, marriage often takes them out of the labour market. Circumstances vary between people from different ethnic backgrounds, and participation and achievement levels vary again with gender and age:

> It is not adequate any more, and quite justifiably so, to refer, for research and analytical purposes, to 'minority ethnic' populations as homogenous. In the past there were the main distinctions between 'British' and 'ethnic minorities' groups. More recently, there has been a gradual realisation that categorisation of populations in terms of ethnic background has to be more complex if it is going to be meaningful. This complexity has to include not only references to ethnic origin, but also to gender, religion and even social class in order to enrich our understanding of British society.

> Traditionally, women in general and women from the working class and minority ethnic backgrounds in particular, were largely, although not entirely, marginalised in research, both as 'subjects' and contributors to its development. It has been a similar case recently for women from the Asian

community and particularly for specific groups more than others, as for example for Bangladeshi and Pathan women, as well as for female political refugees, economic and/or 'illegal' immigrants. Such marginalisation, often unintentional, has serious ethical and practical consequence for all these groups. Despite being the subjects of social policy and heavy government intervention, these minority voices and opinions are unheard, largely misunderstood, misinterpreted or even ignored. This can have serious consequences on the effectiveness of the project and whether the intended recipients take ownership of the initiative.
(Ahmed et al., 2001)

In exploring the relationship between transport, ethnicity and social exclusion, it is important not to draw false boundaries around the experience and membership of ethnic communities: ethnic communities are multifaceted and have relational overlaps, connections and integration into other societal streams. In the research conducted by Oxford University, there was evidence of significant family relationships which crossed conventional boundaries but equally importantly there was evidence of 'neighbouring' practices which also crossed ethnic lines:

I have a very nice neighbour, an English lady, she rings me and says 'What do you want from the shop?' and then she brings things and takes the money. This is for bread and milk and things, not the Indian things from St Marks Rd. On Saturday what she does is she takes me with her. Whatever I want to buy, she puts in trolley. Very very kind lady. She helps me a lot. For vegetables and spices, we keep saying to driver (of social club minibus) that we want to go to shops and he say OK.
(Asian woman in her eighties in Easton, Bristol)

3.2   The best of the literature: accessing unpublished reports

Our research supports the view of the recent Annex B to the DETR's 'National Strategy for Neighbourhood Renewal' which states "There is a significant lack of data about ethnic minority groups".
(Beuret et al., 2000:13)

There is even less about visible religious minorities. Furthermore, when it comes to the transport needs of such groups there is an even greater lack of both information and awareness of the need for information. Thus, for example, the "New Deal for Communities: Race Equality Guidance" which looks at race equality in relation to the four priority outcome areas of jobs, education, health and crime, does not mention transport. (Our emphasis.)
(Beuret et al., 2000:13)

In undertaking the literature review for this book, and the DfT project of which this book is a product, there was a dearth of material on ethnicity and transport. The literature that existed was primarily focused upon personal security, very often in the context of gender, with very little literature attuned to the accessibility and mobility issues necessary to the obtaining of mainstream services by the ethnic population. Recently, and in many ways by chance, access was obtained to an unpublished report by Beuret, Aslam, Gross, Osman and Khan (2000) undertaken for the DfT which lists and summarises a number of other unpublished studies. The report contains no injunction against quotation and is perhaps the most useful piece of evidence in the whole of the literature. We use summary materials from this research to provide evidence of the richness of the tapestry of ethnic interaction with transport albeit largely officially unrecorded – this represents the beginning of a useful policy meeting ground on transport and ethnicity.

Table 3.1   The Relationship Between Transport and Ethnicity: A Summary Table

Study	Key finding
Published Studies	
GLC women on the move. Black Afro- Caribbean and Asian women. Volume 5 (1985)	This study found that Black Afro- Caribbean women's usage of public transport was different to usage by Asian women. Although both groups relied heavily on buses, the percentage of Asian women who used buses was only 42% compared to 79% by Afro- Caribbean women. Asian women were also more reluctant to use the London Underground. It was also noted that Asian women were more likely than other women to be car passengers. During the day time both groups felt safe walking. However at nightime Asian women felt more unsafe to walk than Afro-Caribbean women.
Personal Security Issues in Pedestrian Journeys. DETR (1999)	This Report showed that contrary to the conventional wisdom, in some areas with a high proportion of ethnic minority groups, for example Highfields in Leicester, there was confidence and willingness to walk even though people were realistic about crime rates. The reasons were familiarity with the area, the reassurance of community support and the fact that people had lived in a specific area for a length of time themselves and had not experienced any incidents of physical or violent attacks. It is interesting that this finding contradicts other findings in the same study. For example, in the West Midlands ethnic minority groups felt more unsafe than the local white population. (It may be that the circumstances in which members of ethnic minorities feel safe when walking depends on a variety of factors including the extent of local community activities and/or the proportion of minority residents. Is there a 'critical mass' phenomena?)

Crime Concern & TTR 'Perceptions of Safety from Crime on Public Transport' (1997).	People from black and minority ethnic communities reported higher levels of fear than non minority groups. This was partly due to the higher levels of harassment they experienced.
Somali Women in London by Duale, A.K. (1999)	A study by the Women's Design Service referred to the difficulties experienced by strangers adapting to new environments (for example, Somali women refugees in London).
"So What's the Point of Telling Anyone? – A black community perspective on crime in the Leicester City Challenge area" The Black Community Safety Project, Saini (1997)	This study showed that in spite of a desire to use public transport, people who did not speak English were deterred to use it due to lack of confidence and embarrassment due to language difficulties.

Research Studies

Leicester City Council Mobility Management Survey, 1999	This study of the travel to work patterns of employees of Leicester City Council showed that ethnic minority employees were more likely to use public transport than white employees.
Warwick University – "Analysis of community patterns by ethnic group working paper 1" 1991	Based on an analysis of the 1991 Census travel to work data the following key points were shown: there was a larger than average proportion of the Chinese group that worked from home; similarly Pakistani and Bangladeshi groups tended to be more concentrated in the shorter commuting distance categories than other groups.
CENTRO- "An investigation into ethnic minority usage of transport services for people with mobility problems in Sandwell" (1994)	The aim of this study was to discover why services like community transport had a low usage from ethnic minorities. The findings showed that: amongst ethnic minority groups, it was Asian people who in the main had not heard of services like Ring and Ride. Black and Asian elders were more likely to struggle on to buses, rather than use the services that are provided for people with mobility problems such as Ring & Ride. The key recommendation made to improve the service for ethnic minority groups included making the group more confident and recruiting directly within the minority communities. This meant having more ethnic minority staff present on these services. It was also recommended that better liason needed to be done with community groups as this would be a good way to promote the services.

Glasgow PTE "Towards extending the Strathclyde elderly forum network to the Black and ethnic minority elderly community in Glasgow" (1999)	This was a general study carried out to explore the various social needs (housing, health, leisure, transport) of ethnic minority elderly people. Of the sample, 49% said they either had their own cars or had access to family cars, but this varied between different ethnic groups and was highest amongst Pakistani elders (63%) and lowest amongst the Chinese elders (25%). Public transport was used by 55% of the sample, but again this varied from 57% of the Pakistani group to 33% of the Bengali elders. Some people did not use public transport because they either preferred to walk or felt "apprehensive" about using it.
University of Strathclyde Report for The Strathclyde Elderly Forum by G. Zawdie & N. Riddell, David Livingstone Institute (1999)	This study of 119 black and ethnic minority elderly in Strathclyde showed that although most were entitled to travel allowance, most were unaware of the entitlement. The research also indicated that Chinese elderly had less access to cars (mainly via younger family members) than other ethnic minority groups or the white elderly in the sample.
Nottingham City Council: City Centre Perception Study: Appendix 2 – The Views of Black People – sub report by David Weaver (1995)	This study showed that the demand for travel to the city centre by ethnic minority groups was constrained due to low incomes.

(*Source*: Beuret et al., 2000: 7)

Beuret et al. (2000: 12) alert us to a substantial policy literature on ethnicity and transport which does not surface either in the journals or on the internet. There are many small consultancies and local authorities involved in collecting local materials on transport and ethnicity which are not being integrated into an accumulating literature or body of knowledge which is readily accessible. To give an indication of the extent of these activities on the part of one consultancy – Social Research Associates – we reproduce the table presented in Beuret et al. (2000: 12)

Table 3.2   Issues from Past Research Projects Carried Out by Social Research Associates

Study	Finding
The Meadows, Nottingham (2000)	This study investigated the effect of replacing circuitous bus services within a housing estate by more frequent services on the perimeter. The disadvantages were greatest for people who were disabled or concerned about personal security and this applied to all communities on the estate including minority groups. In addition, one of the reduced services which had provided a link with a specialist Asian shopping area at Hyson Green had a greater impact on this minority group than other residents.
A Strategy for the Provision of Public Transport Information by Nottingham City Council (2000)	This study showed that there was a need for better telephone information services about public transport in minority languages – at least some of the time.
Nottingham City Centre Perception Study (1995)	Minority groups throughout the city needed to travel to a particular suburban area for specialist food and other goods and thus found public transport based on radial routes to the city centre inconvenient.
Leicester SRB5 Childcare Strategy (2000)	Ethnic Minority single parents who were rehoused in an area remote from their community areas needed to travel into the city centre and out again – making two bus journeys – in order to access specialist shopping, religious centres and other community facilities. Many could not afford the double fares.
Merseyside Northern Corridor Study (1995)	Young black school children were reluctant to use public transport due to bullying and violence from gangs.
Blackburn Taxi Study (1994)	Asian women shift workers waiting for early buses were often ignored by drivers and left behind as the bus drove past without stopping. The women believed this was due to racism. The unreliability of off-peak bus services also resulted in discrimination in employment to people (predominantly women) without access to cars.

Nottingham Parking Study (1996)	Muslims visiting the mosque on Friday complained that they had to pay high rates to park next to the mosque, whereas most Christian worshippers parked free on Sundays.
Crime on the London Underground (1991)	Interviews with pickpockets and men convicted of violent attacks on the London Underground showed that they often targeted minority groups as more vulnerable and less likely to receive help if attacked.
Leicester Cycling Study (1993)	Cultural attitudes resulted in women from the Asian community being discouraged from cycling.
London Transport Driver Recruitment Report and other bus company studies (1990-97)	Black bus crews perceived that they experienced more complaints from and attacks by passengers. This was not monitored so it was difficult to check and training to overcome customer care problems was generally not available or inappropriate.
Birmingham Pedestrian Access to Bus stops Study (1999)	In areas with high proportions of ethnic groups, bus stop locations were inappropriately based on outdated patterns of community movement.
Barnsley Public Transport Study (1993)	Religious and cultural beliefs resulted in seating difficulties on public transport in mixed sex environments.
Aberdeen Bus Study (1994)	Telephone Information Services and the structure of fares disadvantaged non-English speakers and those with larger families.
Study of Unlicensed Driving (1999)	Our scoping study indicated higher rates of unlicensed driving amongst ethnic minority groups, including women.
Attitudes of the Asian community towards sustainable living styles (1993)	This study explored the link between Asian ethical beliefs and 'green' issues, with a particular view to the development of publicity for encouraging sustainable life styles. The results showed higher priorities in relation to cleanliness.
Walsall Bus study (1996)	Poor English language skills by some Asian transport staff resulted in many problems for passengers including Asian passengers who spoke a different Asian language.

(*Source*: Beuret et al., 2000: 12)

In the following sections of this chapter, we will be returning to some of these findings catalogued by Beuret et al. (2000) and reinforcing and extending these understandings with the voices of respondents in the Oxford University study.

3.3   The need to adjust travel models: the benefits of cultural awareness

Recent reviews of both the American travel behaviour literature (Batelle, 2000) and the British travel behaviour literature (Beuret, 2000) have pointed up the dearth of culturally aware research in this area. For the moment, we want to turn to the United States merely to observe the development of a new and more vigorous literature on transport and ethnicity: recent work still bears the instruction 'not for quotation' but is available at http://www.fhwa.dot.gov/ohim/trvpatns.pdf and is discussed under the heading Travel Patterns of People of Color. In the United States, transport itself has a place as part of the civil rights movement's history with the famous bus boycott of Montgomery, Alabama (Green, Grieco and Holmes, 2002 @ http://www.geocities.com/the_odyssey_group/ archivingsocialpractice.html). This legendary history on ethnicity and transport has been captured and moved forward in recent moves by Harvard University (http://www.civilrightsproject.harvard.edu/ research/transportation/call_trans.php) to develop a programme which systematically explores the links between transport and black civil rights: [1]

> Americans are increasingly mobile and ever more reliant on automobiles for meeting their travel needs, largely due to transportation policies adopted after World War II that emphasized highway development over public transportation. These and other transportation policies have had inequitable effects on minority and low-income populations, often restricting their ability to access social and economic opportunities, including job opportunities, education, health care services, places of worship, and other places such as grocery stores. Transportation policies limit access to opportunities through direct effects, such as inequitable costs, and indirect effects, such as residential segregation. The indirect effects are caused in part by the combined effects of transportation policies and land use practices.
>
> This report identifies surface transportation policies' inequitable effects. It examines existing research in the area and highlights the critical need for more research and data collection related to the impact of transportation policies on minority and low-income communities. It also makes recommendations to address the racial injustices created by transportation policies.
> *(Sanchez, T. et al., 2003 @ http://www.civilrightsproject.harvard.edu/ research/transportation/call_trans.php)*

The traditional approach to travel behaviour was ungendered. It paid little attention to household scheduling and decision making. It focused primarily on the single purpose work based journey along commuting corridors into city centres on weekdays and on the return journey during peak hours. Within the literature there have been important developments: the work of the Oxford Transport Studies Unit on household activity models (HATS) is regarded as seminal (Pas, 1996) and usefully introduces the concept of household scheduling, bargaining and decisionmaking around journeymaking and household travel resources:

> The activity-based approach to travel demand analysis and modeling traces its roots to the seminal work on urban travel demand analysis undertaken in the mid to late 1970s at the Transport Studies Unit (TSU) at Oxford University under the leadership of Ian Heggie, working under a grant from the Social Sciences Research Council (Jones, et. al., 1983).
> *(Pas, 1996 http://tmip.fhwa.dot.gov/clearinghouse/docs/abtf/pas.stm)*

This model of household bargaining and recognition of the impact of the journey pattern of one household member on the options available for journeying of another could very usefully be integrated into an understanding of ethnic journey patterns. The HATS approach did not assume that all household members had the same travel and transport characteristics: it was understood within this approach that gender roles played an important part in determining journey patterns. However, the HATS approach has not been systematically applied within an ethnic context. There is no main body of research by these authors related to the study of ethnic travel behaviour within any major UK municipality. Adopting the HATS approach would have provided a detailed account of travel decision making within the range of ethnic arrangements and alerted policy agencies to issues such as the relationship between transport, seclusion and the health of women.[2]

The impact of this methodological gap on ethnic travel behaviour has a direct impact on policy: the Mayor of London's Transport Strategy is accompanied by a bibliography which indicates the literature upon which the strategy is predicated and, despite the scale of ethnic diversity in London in a lengthy bibliography only one item on ethnic travel behaviour is included (2003 @ http://www.london.gov.uk/approot/ mayor/strategies/ transport/pdf/final_bib.pdf). Within the report itself, ethnicity receives only the most minor of mentions although those mentions, one of which is provided at the top of this chapter, indicate the seriousness of the problem. Yet the scale of the problem is already evident within the public domain:

> London is one of the most diverse capital cities in the world with nearly 1 in 3 of Londoners coming from an ethnic minority background and with over 300 languages spoken. ...

> (BMECCC report that):BME communities experience a disproportionate risk of victimisation compared to their white counterparts ...Black people have a greater risk of being the victims of a burglary, car crime or violence ...The risk of being mugged is three times higher for BME Groups than for white groups
> *(Black and Minority Ethnic Communities Cracking Crime (Home Office sponsored), 2002 @ http://www.go-london.gov.uk/bmeccc/report.html)*

Given our interest in demand management measures and the impact on social exclusion and given the primacy of London in adopting congestion charging, this lack of sensitivity to ethnic travel issues is disconcerting: in a chapter entitled 'Streets for all' (http://www.london.gov.uk/approot/mayor/ strategies/transport/pdf/4Gstreet.pdf), there is no mention whatsoever of ethnicity. This approach within a leading municipal transport strategy is problematic given the evidence of Crime and Disorder audits in the United Kingdom on ethnic personal security in the wider transport environment (Nottingham Crime Audit @ http://www.nottinghamcity.gov.uk/coun/ department/chief_execs/policy/povertyprofile/08.htm ; Bristol Crime Audit @ http://www.crimebristol.org.uk/): there is no evidence of any recent audit or research into ethnic travel patterns within the Mayor of London's document on Transport Strategy.

New approaches, however, can be found which assist the policy maker in charting this complex terrain, in particular, the development of GIS (Geographical Information System) and GPS (Global Positioning System) systems permit the mapping and capturing of ethnic travel patterns (McCray et al., 2003). Kwan (2003) provides a demonstration of new geographical approaches in her geovisualisation of the differences between Asian and African travel behaviour in Portland, Oregon:

> Finally, the 3D space-time paths of the African and Asian Americans in the sample are generated and added to the 3D scene. These procedures finally created the scene shown in Figure 7 …The overall pattern of the space-time paths for these two groups shown in Figure 7 indicates heavy concentration of day-time activities in and around downtown Portland. Using the interactive visualization capabilities of the 3D GIS, it can be seen that many individuals of these two ethnic groups work in downtown Portland and undertake a considerable amount of their non-employment activities in areas within and east of the area. Space-time paths for individuals who undertook several non-employment activities in a sequence within a single day tend to be more fragmented than those who have long work hours during the day. Further, ethnic differences in the spatial distribution of workplace are observed using the interactive capabilities provided by the geovisualization environment. The space-time paths of Asian Americans are more spatially scattered throughout the area than those of the African Americans, whose work and non-employment activities are largely concentrated in the east side of the metropolitan region. This seems to suggest that racial segregation may involve dimensions other than residential segregation since it may have a significant restrictive effect on the activity space of specific minority groups.
> *(Kwan, 2003:17 @ http://www.ivt.baum.ethz.ch/allgemein/ pdf/kwan.pdf)*

Kwan indicates for us the significance of interrogating existing data sets for their information on ethnicity: such activity conducted on the many travel data sets which exist across the United Kingdom in the many transport institutes and government offices – which are currently not in a common pool – could provide a useful profile for policy purposes. Whilst Kwan is offering us the state of the art geovisualisation of ethnic differences in travel behaviour in the United States, within the UK policy environment even the most preliminary of recording of difference is still awaited. The signs are, however, that US developments in the state of the art will have their consequences for an improvement in UK travel and transport methodologies: the Kwan paper itself was presented at a recent European conference.

In discussing cultural awareness and its benefits in reducing social exclusion in the transport environment, it is important to pay attention to the transport legacy that new migrants into Britain bring with them. For example, experiences of Jamaican transport policy condition the 'voices' of the Jamaican males interviewed within this research. In Jamaica in the 1970s there were tight restrictions placed on the importation of motor vehicles as an economic measure with the consequence that demand management was effectively in place (and which were liberalized in 1991: evidence @ http://www.mfaft.gov.jm/jod/Vehicles.htm) with an associated reduction in congestion. One of our Jamaican respondents in Nottingham, with this legacy of travel awareness in his own experience, asked why the government simply did not constrain the supply of vehicles – an understandable question given his transport history. Similarly, and more widely spread, was the West Indian distrust of public transport, a form of cultural path dependency: a legacy which is the outcome of the highly dangerous circumstances of public transport in Jamaica and unreliability elsewhere in the Caribbean. This legacy results in the aspiration of owning a car as quickly as possible amongst West Indian men. Cultural awareness provides an understanding of two different relationships to demand management within the same community.

3.4 Learning the journey through listening: ethnic voices, ethnic frameworks

In researching, the relationship between transport, ethnicity and social exclusion, it became abundantly clear that there were complexities around journeying at every stage of the process. It is equally clear that these complexities are under-discussed in the policy literature and hidden from the transport professionals by the lack of cultural awareness currently contained within professional training with some notable exceptions:

> Concerns about crime while travelling can deter people from walking, cycling or using public transport. This may be a particular problem in more deprived areas. For example, people in the most deprived areas are around five times more likely to say that they are concerned about crime in their area and safety at bus stops than those in the least deprived areas.

Crime and anti-social behaviour affects both victims of crime and witnesses to such incidents. It can also lead to other problems such as bus operators removing or re-routing services.

Restricted access to suitable transport can contribute to social exclusion by restricting job and learning opportunities and access to healthcare, food shopping and other local activities. Access to public transport is particularly important in deprived neighbourhoods since among the 20% of households with the lowest incomes, 63% do not have access to a car.

Certain groups are more reliant on public transport than others. Research has shown that women from black and minority ethnic communities are more dependent upon public transport than other groups. Women typically make more journeys by bus and on foot than men and travel at off-peak times more often than men. Furthermore, many older people rely upon public transport to maintain their independence.

It is, therefore, important to consider the safety, and concerns about safety, of people using various forms of public transport such as buses, trains, taxis, trams and the London Underground. It includes tram and bus stops or shelters; stations; and routes to, from or between the start and end points of journeys by public transport. It also includes journeys by foot and bicycle. *(renewal.net, Crime and Transport, 2002 @ http://www.renewal.net/Search.asp)*

NACRO has recently drawn attention to the fear of crime in the public transport environment and noted that it has a strong ethnic dimension:

Fear of crime can deter people from using public transport and some groups are particularly affected. Black and minority ethnic people's fear of crime is higher than that of white people, some women will not travel after dark, and parents restrict their children's usage. A DETR survey found that fear of crime while waiting for a train or bus after dark is greater for women than for men, with bus stops being considered less frightening than railway platforms. In the survey 44 per cent of women and 19 per cent of men felt unsafe waiting for a bus, and 53 per of women and 23 per of men felt unsafe on a railway platform after dark. However, both men and women feel safer once they have boarded their bus or train, with buses again rating as less frightening than trains.
(NACRO, 2003)

Research in the United States confirms how threatening the public transport environment can be for people from ethnic communities: McCray (2000), reporting on a study of 20 low-income women in the City of Detroit between the ages of 18 and 46 (eleven of whom were African American, two of whom were white and seven of whom were Hispanic) states that:

The majority of participants reported being within walking distance to bus stops. However, the environment around the bus stops was considered to be dangerous.

(Researcher)TM: "Is it safe to take the bus?"

(Focus group participant) M1: "No it's not! Not at all! I live off the west side. Grand River is one of the worst bus stops that you can be at. You will get robbed! My sister-in-law, they robbed her for a dollar at the Grand River bus stop. It's the Plymouth, River, Greenfield [bus routes]."

(Focus group participant) M2: "Well yeah, I'd have to say I've been robbed off of Greenfield."

(Focus group participant) M1: "I know. Plymouth, Greenfield is one of the worst routes. Plymouth, Greenfield, and Grand River are the worst, and I live by all three of them!"

(Focus group participant) M3: "And Finkel [bus stop]. My mother got robbed, and they didn't even want anything. They just wanted to assault her. They hit her with poles. She gave them the purse, but they didn't take it. It was cold."

Our own evidence supports the perceptions of substantial levels of fear around the use of public transport within ethnic communities. Consequently, this section breaks journeying by public transport down into its component parts and provides the opportunity to learn the shape of an ethnic journey on public transport through listening to ethnic voices. It also contains suggestion from these voices for improvements in the present public transport framework.

Organising a trip:

The assumption current in the travel behaviour literature is that individuals freely determine whether to make a trip and where to make that trip. In general, information is not problematised but it is assumed that information is readily available. In practice, there are significant limitations on who can make a trip, where to and with whom. At the start of this chapter, a respondent's voice was used to demonstrate that issues of seclusion are relevant to ethnic travel behaviour within Britain. This is of consequence for both genders: as the quotation reveals, the social imposition of seclusion upon women entails that tasks which are usually undertaken by women have now to be undertaken by men or by women of a different ethnic group. Immediately, in this brief insight into ethnic household organization, we are forced to an understanding of the complexities of travel decision making.

Of course the complexity of household decision making in relation to travel is not only a property of ethnic minority households: bargaining over access to travel resources occurs within all households. Our research into gender and lifecycle groups, a part of the same study, revealed bargaining for the use of the car in a Bristol low income household where the dominant driver was the wife:

I tend to use the car (my wife's) if I urgently need to make it to an appointment. Otherwise I mainly walk. I sold my own car a few months ago as it hadn't been used for 3 months. If the car breaks down, I call the AA.
(Retired white man in mid 60s, with working wife, Southville, Bristol).

The complexity of bargaining around travel decisions can be more complex in household where there are more complex cultural rules around travel rules and roles. The evidence from Beuret at al (2000:42) is that the consequence of these rules and roles is that Muslim families tend to travel as a group: this meets the formal Islamic religious rules around female travel (Islam on line @ http://www.islam-online.net/English/hajj/2001/WomenandHajj/article1.shtml) and female roles in society. Beuret et al. (2000) note the failure to meet the travel needs of any part of the family group can impinge upon the travel opportunities of the other members:

> given the tendency of Muslim families to travel as a group, not meeting the needs of some members (who are especially likely to be the female members) could deter the others.
> *(Beuret et al. 2000:24)*

Where cultural norms around travel have been predicated in the communal behaviour of the extended family, widowhood can come as a severe deprivation:

> Sometimes you can't find anyone to do anything for you. Everybody is working in this country so how can they do anything for you? You have to struggle by yourself by bus or taxi.
> *(Widowed Asian woman, 60 years old, Easton, Bristol)*

In such circumstances, organizing a trip is far from an easy experience: routes are unknown, familiarity with fare systems is weak and negotiating the public transport system represents a substantial barrier. Even for younger people, the obstacles are serious:

> I'd like to say that they have forced people to use their cars due to the trams being introduced because the bus stops have been moved. I'm a car user most of the time but when I went to town by bus, all the stops had been moved and you don't know where to go. You find then that people don't want to search for a stop and so they take their car next time.
> *(Asian woman in her twenties, Hyson Green, Nottingham)*

The organising of trips by older persons clearly owns its fair share of difficulties, difficulties which are compounded by ethnic dimensions, but the organizing of trips for younger persons holds an equal share of difficulty:

It's very funny in Bristol, the local authority will guarantee you education but not a school near where you live and they talk about a walking distance being 3 miles. So you're expected to walk with a young child 3 miles (and over 3 miles) then they'll give you a pass. And then the bus company won't issue return tickets before a certain time.
(Asian man, 35 years old, Easton, Bristol)

Overall, participants in the Oxford University research reported that difficulties in making a journey can be faced before the trip has even begun.

Waiting for the bus:

Waiting for the bus is not a simple business for many of our ethnic respondents. In terms of their information context, bus stops are information-unfriendly places. Timetables are often not available, even where timetable information is available it is generally available either in a language form or a sign form which creates a barrier to understanding. New electronic signing at bus stops (such as Countdown messaging) is rarely available to ethnic minorities in our study, however, the travel awareness of this group embraced the knowledge that, whilst it was not available to them, it was available to others. The issue of minority languages and real time electronic signing at bus stops requires addressing as does the geographical distribution of the high end public transport in route information services: the digital divide can happen even at a bus stop. Discussions addressing social exclusion issues through the improvement of public transport have engaged with the potential for new information communication technologies to improve the quality of the public transport experience: the potential for real time information systems is recognized, the ethnic dimension of this need has been visibly neglected:

GPS technology is helping to deliver route reliability and provide real time information, smartcard ticketing offers both discounts and faster journeys, operators are wakening up to the need to provide comfortable, reliable services at turn up and go timetables.
(Commission for Integrated Transport: http://www.cfit.gov.uk/pn/021202/)

Policy discussions remain blind to ethnic issues in a context where the lived experience of poor information is recordable but, as of yet, largely unrecorded:

Very often you get buses come down and the first one has no seats but it takes on passengers and the others are empty but they won't take on anyone. Seems to me rather pointless, a lot of people say they won't bother and they just go home when they can't get a seat. That's really hard especially for old people. And the only place where they have the information system on the bus stops is in the more respectable part of town in Clifton.
(White man in his 70s living in ethnically mixed area of Bristol)

I think the biggest problem is a lack of information about use of transport –
about what's available and what's not available.
(Asian man, 50 years old, Easton, Bristol)

There's no transport information for new arrivals, new refugees. I think it is
all in English, I've never seen it in another language. Also this is the only
way the refugees can get around, they can't afford cars and passes. They end
up feeling very isolated and they are forced to walk and they can't get to
work.
(Black African man, 35 years old, Easton, Bristol)

Given the difficulties described by the Oxford University study participants
from Bristol, it is interesting to find the city being presented as at the
leading edge in terms of public transport information provision (Transport
2000 Trust, 2003: 58). The contrast between the lived experience of
residents of inner city ethnic areas in the Oxford University research of lack
of bus information and the services being described below in the area of
Clifton (one of the least deprived wards) is clear:

Bristol is developing a comprehensive information strategy. This includes
real-time information at a growing number of bus stops and a door-to-door
bus journey planner, which allows users to plan bus journeys via the Internet
and at stand-alone high street kiosks. At the Clifton Down shopping center a
Travel Advice video wall has been installed. This can show up to ten
different windows at one time, providing a range of real-time data for bus
and rail services, together with air pollution levels and promotion of transport
initiatives and campaigns. The video wall helps transform the open foyer
area into a departure lounge. The public can relax in a comfortable, warm
environment and monitor accurate, integrated public transport information
covering all modes serving the shopping center.
(Transport 2000 Trust, 2003: 58)

I live in Staplehill Road and when I catch the bus, there is no certain time.
Yesterday I left at half past 8 and got home at 1030, only one bus.
(Black African Muslim woman, 30 years old, Fishponds)

I was going to say the buses are unreliable especially on Stapleton Road. No
bus for 45 minutes and then you get 5 buses at one time.
(White woman living in ethnically mixed neighbourhood, Easton)

There are no timetables and if there are they are vandalized.
(Black African Muslim woman, 30 years old, Fishponds)

Sometimes you have to catch 2 buses and if you miss one you miss them all.
(Asian woman, 45 years old, Easton)

Maybe it would be useful if some kind of system was put in the shelters
which would tell you when the bus was coming.
(Asian man, 50 years old, Easton)

There's a lot of those information systems in Clifton.
(Asian man, 40 years old, Easton)

The fact that participants in our research knew about the superior public transport service provision in Clifton makes their experience of inequity in the transport environment even more pronounced:

They don't really use the buses in Clifton and they've even got the gas bus.
(Elderly white woman, Easton, Bristol)

For years they had the best buses, the poshest buses and the best service when a majority of people in Clifton would never be seen on a bus. And they are not far from the city centre anyway. Catching a bus through St Pauls is hell. You could wait hours for a number 5 – it's notorious for not turning up. Late night services are terrible too.
(White mother of Black Caribbean daughter, 40 years old, Easton, Bristol)

They can't provide an adequate service during the daytime and now they've come up with this all night service for people that go clubbing and that at night. I think they run it on the hour and on the half hour from town but it will only go to certain points. It's not like a bus route that goes where the other buses go. It doesn't go to St Paul's. They go from town to Fishponds on a straight, or town to Clifton but not to the inner city.
(Black Caribbean teenage girl, Easton, Bristol)

It would be nice too to know if you are at a bus stop when the bus is coming, like you know in Clifton – it says number so and so will be here in that number of minutes.
(Elderly white woman, Windmill Hill)

Not only is information absent at the bus stop but bus stops themselves are often absent or not operational.

Also the park and ride bus shelters and the ones for the train station have live information but we don't have them, and half our bus stops have no shelters at all.
(Black Caribbean woman, 20 years old, Easton, Bristol)

I live in the biggest block of flats in Bristol but there's no bus stop on that street. It's a good 10 minutes walk away.
(White elderly woman, Windmill Hill, Bristol)

Sometimes you have to stand in rain waiting for the bus and snow and that's why car is very important.
(Asian woman, 40 years old, Easton, Bristol)

Even if you make it to the bus stop, it's so long to wait so you don't go anywhere and there are very few bus stops with seats where elderly can sit anyway.
(White elderly woman, Windmill Hill, Bristol)

And even where bus stops are available and information is available, the social context of the bus stop can be alarming and threatening.

At the end of my road, there are prostitutes and I hate to see young girls, especially young women standing at the corner of my street at the bus stop because they now stand at the bus stop picking up men. Even I hate to stand on the corner of the road because people are looking at you and the way they are looking at you, you think 'Oh my God'. This is especially in Easton and St Paul's. They are all down by Sussex Place, down where I live and all up Fishponds Road. So it is difficult to stand at a bus stop at all even in the day time now.
(White woman in mixed marriage, 40 years old, Easton, Bristol)

I was at a bus stop near the shop and there was elderly man in there too. These kids of about 10 or 12 kept riding round and round and banging the bus shelter and it was really terrorizing that gentleman.
(Asian man, 50 years old, Old Meadows, Nottingham)

Figure 3.1 Waiting for the bus on Stapleton Road, Easton, Bristol

Listening to ethnic accounts of waiting for the bus brings us to the reliability of buses, the frequency of buses and the variabilities in service provision by time of day or by position in the week. Glenn Lyons in his inaugural lecture (http://www.transport.uwe.ac.uk/LYONS-Inaugural.pdf) reports:

> Transport Trends (ONS, 2003) argues that "the availability of bus services is fairly good overall" on the grounds that nearly 90 per cent of households in Britain live within 13 minutes' walk of a bus stop with a service at least once an hour.

Many of our ethnic respondents and their white neighbours are reporting situations and residential circumstances which do not meet these standards.

> Whatever they put in place, it needs to be reliable. People going to work, people going to hospital, you shouldn't have to leave home an hour and a half before your appointment just to make sure you get there on time.
> *(White woman, 58 years old, Easton, Bristol)*

> Especially when you got hospital appointment, you have to leave house one and half hour before to wait for bus. If you catch it in time you lucky, if not one and a half hour on the wait.
> *(Asian woman, 60 years old, Easton Bristol)*

Furthermore, many of our respondents report that their journeys require interchanges to get them to the services they require: and where there is an element of discretion, forced interchange may result in the suppression of the trip:

> Sometimes I don't go to a meeting if it involves 2 bus journeys.
> *(Retired Southville, Bristol resident)*

'Interchange' necessarily means at least two waiting periods for one bus journey and one of these waiting periods necessarily happens in an area where the traveler has no place of sanctuary. Experiencing danger at a location where it is easy to return home is very different to experiencing danger in a location where no rapid exit journey can be made. The appropriate measure of a good overall coverage would be one in which households were living within **thirteen minutes of a bus stop which provided them with a direct service** to necessary social, economic and health destinations. Many ethnic households are reporting on high levels of indirect journeys:

> I would like to see that buses are pointed in direction that people want to travel and to see more fisticated buses even if you have to pay a little bit more.
> *(Black Caribbean man, 45 years old, Easton, Bristol)*

The findings of Cass et al. (2003) support the views expressed by participants in our study. Reporting on work done in the CHIME study for the DfT, Cass et al. highlight the difficulties involved in 'trying to represent a moving landscape of public transport opportunity' and point to the need to look for 'simple methods with which to represent the temporal and spatial characteristics of public transport systems'. The authors note that:

> There is a significant gap between the experience of potential 'users' wanting to make specific journeys and the representation of public transport systems in terms of the 'density' of buses per hour or access to bus stops.

On the buses:

Using the bus is far from an easy business for many of our ethnic respondents: within our evidence, older ethnic women are involved in the childcare of their grandchildren and experience the difficulties of boarding the bus with their own limited agility accompanied by grandchildren, pushchair and the bags and packages necessary for making a journey with a child:

> It is very bad in the morning because my daughter lives in Fishponds I used to take a bus with a pushchair with her children. So much struggle you know, so much rush in the morning and in the afternoon as well, rather you can take taxi but taxi's so expensive. So what can you do?
> *(Asian woman 60 years old, Easton, Bristol)*

Many young mothers report difficulties taking their prams onto public transport: even where there is a preference for public transport use, it is frequently impossible for a mother with pram and child to make the journey in this way. Often such journeys are made to obtain shopping with the pram providing a form of transport for heavy items. Under such circumstances, alternative car based solutions are sought – in the case of one of our ethnic respondents, this involved making arrangements with her boyfriend who lived in a separate accommodation to drive her on her trip:

> I have to rely on my boyfriend to take me shopping. If the bus would let me on with a pram, I could take a few bags home then but I have to walk and it does seem quite far when you're walking but on a bus it doesn't seem far.
> *(Black Caribbean single mother 19 years old, Easton, Bristol)*

The routing of buses: all around the houses

The research into transport, social exclusion and ethnicity indicates that it is important to consider ease with which journeys may be made when planning bus services: for ethnic minority communities, social links may be across town necessitating interchange, even the trip into town may require the use of more than one bus and the growth of traffic calming measures on residential estates may have resulted in bus services operating around the periphery of the estate rather than into its core making access to public transport services difficult:

It's expensive too if you've got to go into town like and have to take 2 bus journeys to get anywhere and 2 to get home. Majority of the services you have to change in the centre. There used to be buses all the way through but it was long trips and they've cut them off at the centre and now you have to get another bus. I think we've got the dearest bus service in England in Bristol.
(White woman, 58 years old resident in mixed ethnic neighbourhood, Easton, Bristol)

If you're looking at green issues and you look at transport into local areas, we're not knocking the road humps but the big effect on buses is that they go around the outside, not many go through the middle. So if you have buses going through the heart of the community instead of around it makes better sense in terms of access.
(Asian man, 35 years old, The Meadows, Nottingham)

Beuret et al. (2000) also found that the patterns of ethnic minority travel had not been taken into account when service provisions were being designed:

One problem was that the journeys which ethnic minority groups wanted to make were more likely to involve interchange. This was because destinations were often out of centre to specialised shopping, education or working areas in a particular sector of town. On the other hand timetables were organised around radial routes. We found no examples of timetables tailored to local ethnic minority patterns of travel.
(Beuret et al., 2000:33)

In our own research evidence of ethnic difficulties in access to services both mainstream and culturally specific was apparent:

I used to live in Lenton and you could get everything nearby, services like Post Office, Halal meat shop, medical and dental all nearby, within walking distance, so it was not necessary to keep car in these circumstances. If you wanted to go to town you could also walk. In new area, you can only do one job in one day. Similarly if these facilities are provided at close quarters, then walking will be feasible. To go to anywhere now I have to go on bus and bus is half an hour service, so whole day is wasted. Then at 4 o'clock you have to pay very expensive fare.
(Elderly Asian Muslim man, Hyson Green, Nottingham)

Furthermore, our own research found evidence that changes in land use planning and in bus routings impacted upon the ability of Asian residents to gain access to the ingredients of their traditional diet. There are direct cultural consequences to the changes in routing: in Nottingham, as we shall see later, the discontinuation of through buses impacts on these aspects of life.

Poor weekend services: suffering on sociability

In exploring the relationship between transport and social exclusion in relation to ethnic minorities, it is important to move beyond the analysis of day and peak time services: areas which are adequately served during the day time hours see severe restrictions on public transport availability in the evening hours and at weekends. The work by Beuret et al. (2000) finds that ethnic minorities experience limitations on their sociability due to lack of weekend services:

> The much-reduced weekend services by transport providers was heavily criticised by all groups. It was unanimously felt that both train and bus services were poor at the weekends, especially on Sundays. This had become problematic for members of the public as in recent times Sunday had become a popular shopping day. People generally felt that transport providers had failed to take in to account new patterns of consumption and expenditure.
> *(Beuret et al., 2000:19)*

In the research conducted within the Oxford University study (Rajé, 2003), similar views were expressed and experiences reported:

> What's good? Well I take a bus every day and I think the service is disgusting. It's unreliable, during the week it is alright because it runs between every 5 and 10 minutes but at weekends just forget it. The 25. …And to let buses run more often at weekends, everything is at a standstill at weekends. I mean the frequency is brilliant on the weekday but this time now (after 1800 on a weekday) its going to take me ages and I'm going to have to take a taxi home. You know people do work in the evening, people do go out in the evening.
> *(Black Caribbean woman, 45 years old, St Paul's, Bristol)*

> It's OK for school on a weekday but on a weekend I always go out and I could wait up to an hour for a 25 bus.
> *(Black Caribbean schoolgirl, 15 years old, St Paul's, Bristol)*

> Also the last bus back is quite early about 2315 but half the time they run early so you could miss that. Like by the time you get out of a pub you've missed your bus and then it's a fiver for a taxi. If the last bus was at midnight, you'd have time to get your bus. And they do the night bus on weekends but that just goes to the far areas, never goes anywhere local – like it goes to Ashton, Bedminster. Doesn't pass this way. I think they've got 6 NightFlyers and they go to Ashton, Southmead, Cribbs Causeway, can't get one to St Pauls, can't get one to Easton except one stop on Stapleton Road.
> *(Black Caribbean woman, 20 years old, St Paul's, Bristol)*

Bad thing about this bus service now is I like to go into Nottingham for a meeting in the week which is from 7 until 9 pm. I come out of the meeting just after 9 and I've just missed a bus, next one is at 10. One an hour coming out of town is ridiculous. It's crazy.
(White man, living in ethnically mixed area, the Meadows, Nottingham)

I don't have a car, tend to walk within reasonable distances but I do use the buses. If it's evening, a very small part of the day after working, from 6 til 9, if you want to go somewhere and it's waiting half an hour here and half an hour there, it is biting into that little bit of time you've got. And sometimes I just say no and don't go and meet friends or go to relatives.
(Asian man mid 30s, The Meadows, Nottingham)

Interestingly, there are variations between municipalities in Britain and within municipalities themselves as to the quality of local service provision in off-peak hours as reported by ethnic respondents: a black Caribbean woman who had previously lived in Birmingham and now live in St Paul's in Bristol provides a useful insight into ethnic community unrecorded 'travel awareness':

I find that really weird because originally I come from Birmingham and they have a really good night service. Go out on a Saturday night and you know the buses are running. So when I came down here it was a shock to my system.
(Black Caribbean woman, 45 years old, St Paul's, Bristol)

The preference for bus conductors:

The preference for conductors as a safety measure against on board crime and as assistance for those with mobility difficulties frequently emerges within research into low income public transport needs. Beuret et al. (2000: 22) found that all ethnic groups interviewed within their study reported a preference for the retention and reintroduction of conductors on buses as a personal security measure:

All groups interviewed suggested there needed to be higher levels of staff present on public transport. All groups felt there should be conductors on the evening buses at least, if not on all services. Personal security concerns heightened for all people at night and this meant they did not like to wait for buses on their own at bus stops or at train stations. Many people described 'incidents' such as name calling, jostling and other unpleasantness which were often as upsetting and threatening as actual reportable incidents.
(Beuret et al., 2002: 22)

Within the ethnic neighbourhoods researched in this study, the same preference was indeed evident:

Certain areas that buses don't go after dark for fear of attack like Knowle and Southmead. When they had conductors on the buses at least the driver had some sort of back up. So money could be put into security measures on the

buses. However, people know when there is a sign on a bus saying that it is covered by CCTV that it's not true.
(White woman living in ethnically mixed neighbourhood, 58 years old, Easton, Bristol)

Speaking with the driver:

Beuret et al. (2000:16) found that language could be a barrier to use of public transport:

> Some women, both elderly and younger, did not use buses at all and only went out in the family car accompanied because they could not speak English and therefore communicate with the bus drivers.

McCray (2000) reports similar findings amongst some ethnic groups in the United States:

> The Hispanic women reported they did not take the bus because of the language barrier.
> (Focus group participant) M4: "I don't really know about the bus, where it's going. I have a language barrier, and don't speak English. So if I have a question, how do I ask the bus driver? Where are you going to stop if you don't even know where to stop."

The Oxford University research also found that the lack of spoken English was a barrier to use of public transport: communicating with the bus driver could be fraught with anxiety. These voices below are an extract from a focus group held at the offices of a community translation service in Easton, Bristol and capture the communication problems succinctly:

> *Black African Muslim woman, 24 years old:* When I came new here my husband used to write the name for me of where I go and I show to driver because if I say it he can't understand me.

> *Asian Muslim woman, 32 years old:* Most of the people who don't speak English hesitate to get on a bus because they don't know the rules and you need to talk to driver.

> *Asian Muslim man, 40 years old:* I think a big company like the bus company must make information available and teach drivers to understand people. I still find it difficult to go through times for buses.

Travel awareness: the theory and the practice on the ground

As Beuret et al. (2000) recognize, ethnic minorities often experience additional difficulties in gaining travel information as a consequence of language barriers and lack of access to the locations where information is available. As we have seen, these problems are compounded by lack of familiarity as to how the systems operate and, on occasion, by the unhelpfulness of drivers. In this light, it would seem appropriate for instruction to be provided on how to use the public transport system for people such as some of the Asian elderly who find the car has often been their dominant mode of travel. As they age, some people find that they no longer have access to the car – the driver may die leaving a non-driving spouse, physical problems may prevent driving and economic costs may render driving unfeasible. A need for training on how to use a bus and a need for information to be readily available in their own languages were seen as key to overcoming apprehension about using public transport.

Giving up on public transport: using scarce resources on a taxi

The Oxford University study indicated that people in low income communities are often dependent on taxis for key journeys when conventional bus and demand responsive services do not deliver. For people with constrained personal and family budgets, this expense can often not be justified and the trip is then foregone or a trade-off is made in order to allow the journey to be made:

> Spent all my money this week on transport. I like to walk but if I go shopping I have to take the taxi back. I also take buses and they're OK but I seem to spend all my money getting places. If I have the kids, I take a taxi. Yesterday, I went shopping in town on the bus and then came home by bus. I get very tired at the moment but still try to walk as much as possible.
> *(White heavily-pregnant woman, 28 years old, in ethnically-mixed neighbourhood, Hyson Green, Nottingham)*

> Everybody doesn't have car and can be suffering by bus. It is so difficulty when you miss your appointment or visiting someone. Taxi cost too much.
> *(Asian elderly woman, Easton, Bristol)*

> Aldi's (supermarket) is on the boulevard and I have to get a taxi back from there rather than both ways because I can't afford it both ways and I can't get a bus there.
> *(White elderly woman, Clifton, Nottingham)*

> I find when I wasn't driving to go shopping was a pain. You know to go to Iceland. Tesco's isn't too bad because there's a bus, go to the butcher on Stapleton Rd or Iceland it's a pain because you have to carry it back but I used to walk with the buggy. At the time you never had a delivery service. If you pay for a taxi there and back that's near enough £10 and that's half your shopping and could have been spent on food. It was awkward. Now I have a car, I take my friend with me too.
> *(White woman, 25 years old, reporting on giving lifts to Black Caribbean friend also attending focus group, St. Paul's, Bristol)*

> Sometimes I'd like to go out but unless I've got the money for a taxi, I don't even bother with it.
> *(African-Caribbean woman, 45 years old, St. Paul's, Bristol)*

Also when you go shopping and take a taxi, every bag is 50p. These are the disadvantages for people who can't use public transport and don't have car. *(Asian man, middle-aged, Easton, Bristol)*

There is a need for greater awareness in the policy environment of the types of, often basic, challenges people from ethnic minority groups can face in using the transport system. Even amongst organizations that are lobbying for an improved public transport experience for users such as those on low incomes, older and disabled people, women, families and young children, there is little reference to the needs of ethnic minorities: a word search of a document produced by Transport 2000 Trust (2003) – 'At the leading edge: a public transport good-practice guide' – found the word 'ethnic' only twice. These references related to fear of crime and efforts to employ an ethnic mix at a particular rail operator: there was no call for information in other languages than English, individual travel planning for people who may be unfamiliar with the public transport environment such as new arrivals or older Asian women who had previously been car-dependent.

To conclude this section, all these factors associated with waiting for a bus, traveling on a bus, interchanging within the transport system, experiencing travel uncertainty and personal insecurity in the public transport environment can deter people in ethnic minority communities from using the bus. And our respondents report that they do so deter. There are consequences in this situation for demand management policies. For those members of the ethnic community with sufficient incomes, private cars begin to look very attractive. If demand management policies are introduced without equity audits and accompanying remedial measures such as the development of demand responsive transport, social exclusion of ethnic groups may in fact worsen. For those members of the ethnic community who have highly restricted incomes, journeys may be suppressed, even essential journeys such as health; life choices are affected with opportunities for employment being lost or jobs surrendered or education discontinued and where journeys are made they are made in conditions of stress. The improvement of public transport appears to be a necessary condition for successful demand management policies given the existing circumstances of transport and social exclusion in the United Kingdom (Social Exclusion Unit, 2003).

3.5 Bus boycotts: forgotten ethnic experience

Before leaving this chapter, it would be remiss not to mention the ethnic experience of boycotts in the transport environment. In terms of ethnicity, we tend to be more familiar with boycotts in the US and in Southern Africa (Green et al., 2000), yet boycotts also took place in the UK. Boycotts represent:

the use of transport systems not as a mechanism for extraction and control over populations by dominant political and economic interests but as an organising resource which can be used for resistance and change by the historically vulnerable.
(Grieco, 2003b)

Phillips and Phillips (1998: 226) report that prior to the Race Relations legislation in 1965, 'there was no legal redress for blacks who faced discrimination in employment.' According to Phillips and Phillips, in Nottingham, a Jamaican migrant called Eric Irons

> ...had to mobilize support to negotiate with the City Transport company before black workers were employed. In Bristol desegregating the bus company took a major campaign, which was led by Paul Stephenson, who was not a West Indian, but came from an English, mixed race family.

The authors provide details from an interview with Paul Stephenson which describe the boycott in Bristol. Some of this text is reproduced here with permission from the publishers to capture the experience as closely as possible:

> ...A boycott. Luther King had boycotted the buses in Alabama over black people being forced to stay at the back of buses, so I thought I'd do the same tactic but this time over employment...So I called the boycott, not just for the black people. I said, anyone in Bristol who feels that this policy is wrong, don't use the buses. And it did what I hoped it would do, it caught the imagination, first of all of the local press, who blazoned it in headlines, and it also caught the imagination then of the national press, and it became known as the Bristol bus dispute...

> ...Ian Painty, who was then the director of the bus company, came out publicly and defended the policy. He said, 'White people won't work with black people on the buses and, moreover, it could be that we'd lose even more customers if we had black drivers and black conductors on the buses.'...And one woman actually was heard to say that it would be wrong, morally wrong, to be seen on the bus after nine o'clock with a bus driver who's black...

> ...It took us six months, nevertheless, to get the first black person on, and it happened to be a Punjabi, so I was very pleased...We'd fought the battle, it was a moral battle, it wasn't a political one. And it was the first black-led protest, national protest against racial discrimination and the promotion of equal opportunities that this country had seen...

Grieco (2003b) states that the centrality of transport to modern organisation creates a tool for disruption and civic protest of routine economic business. Transport boycotts have been aimed at reducing revenues of powerful interests by the withholding of fares – a form of consumer boycott – or labour – transport worker strikes – as a pathway towards social change. As we have seen, the Bristol boycott was successful in changing ethnic minority employment policy at the bus company.

3.6   Conclusion: revealing the hidden

In this chapter, we have used cultural practices around seclusion to suggest that there are very hidden dimensions of the relationship between ethnicity, transport and social exclusion which require revealing. Women in strict seclusion have to accomplish activities through the physical mobility of others: the issue of indirect accessibility to services, in such a context, surfaces as highly relevant to the exclusion debate. Even women from some ethnic minority groups who are not in strict seclusion encounter difficulties accessing key services:

> The report (on the largest national survey of the reality of race in Britain) found widespread dissatisfaction among Asian women with a wide range of services "including housing, welfare benefits, police services, childcare, health and mential health specifically, and services for elderly people, children and carers. Many women had limited experience of using any services and had no knowledge of where they could go for advice and information".

> The report highlighted the concern of many Asian women that male members of the household controlled access to advice and information: "This, they felt, was largely down to the fact that women were not often allowed out of the house. One woman explained "I don't have much contact with anyone."
> *(The Observer, 15 July 2001 @ http://observer.guardian.co.uk/focus/ story/0,6903,521949,00.html)*

The journeys not made because of poor evening bus services or poor weekend services must also be considered within the realm of the hidden: sociability and cultural practices are fractured with unmeasured consequences for the quality of life of ethnic residents. The journeys that are unmakeable simply because of the unavailability of public transport, demand that can not be modeled from origins that are determined by the allocation of social housing without regard to accessibility or mobility considerations in respect of ethnic minorities and indeed the wider community of the vulnerable – these pressures on travel and transport are all hidden in the present by the underrepresentation of ethnic voices in travel and transport research:

> I just want to confirm that there are some areas where buses are not available. There's a lot of Somalians that I know that have been offered a house up there in Hawfield but they say they will not go because of the bad crime in that area, it is run down and deprived, but also because there are no buses and so you have to stay there and the gangsters are on the chase. The women are really scared.
> *(Somali male outreach worker, 36, Easton, Bristol)*

The research reported in this book experienced many difficulties in gaining access to ethnic voices on the travel and transport environment. Unlike much of the research undertaken in travel and transport it reported on the difficulties involved in seeking to reach 'hard to reach' groups and hidden travel patterns and barriers within British urban life. Nevertheless 'hard to reach' groups must be consulted in the urban transport research process as these are the very groups likely to experience the most extreme problems – the very factors which result in such groups being 'hidden' contribute to their transport disadvantage and consequent social exclusion.

In the case of our own research as the study progressed, there was an apparent gap in representation from the traditional ethnic minority population of the Black Caribbean community. Similar difficulties were experienced in Nottingham. In Bristol, it was difficult to find community groups amongst this population generally and even more so to find ones who were willing to participate in the research, for example, a flyer was sent out with the Bristol Black Development Agency's regular mail-out to all black and multi-ethnic organizations in the area, asking people to participate in the research but there was not one response; contact with the Jamaican owner of an Internet Café in the St Paul's/Montpelier area to arrange a focus group with young people using his facilities was similarly unsuccessful. In order to explore this community's views and experiences further, and in particular those of the young, contact was made with a youth project on the Lower Ashley Road in St Paul's to arrange a focus group with some of the members – this avenue too proved unsuccessful with the outreach worker being unwilling to assist in organizing interviews (this despite the Caribbean background of the researcher and the fact that the outreach worker had been contacted through a fellow outreach worker). Contacts were also sought amongst nurses' groups – a profession in which ethnic workers are very visible – to see whether a meeting could be held wiith them, not only to capture views from the Black Caribbean community but also to hear about the experiences of shift workers.

Having tried all these channels and since none of these contact pathways were fruitful in generating focus groups, the researchers decided to return to Bristol and try to obtain a greater understanding of the Black Caribbean population's travel patterns and views on transport by a combination of interviews, interview diaries and, if possible, spontaneous focus groups. A spontaneous focus group is one in which members of the target interview group are gathered together at a key activity destination in activities which permit group conversation for an extended period. This led to a focus group meeting being held in a hairdresser's on Stapleton Road, an ethnically mixed neighbourhood of Bristol, and interviews being conducted with people at key activity destinations such as bus stops, medical facilities, pharmacists, shops and post offices – locations in which there were queues and thus time available for conversation. In this way, the voices which are normally hidden or silenced were given space and time for expression.

In Nottingham, several attempts were made to contact groups representing the African-Caribbean community such as ACNA (African-Caribbean Nationals Association) and ACFF (Association of Caribbean Families & Friends) but these were not successful. The Queen's Medical Centre was contacted to arrange focus groups but the management would not allow researchers access to staff since there were on-going negotiations with the unions about parking and other transport issues during the period of research related to preparation for the Work Place Parking Levy. Poignantly, the level of policy activity around parking issues made it difficult to get access to discuss travel, transport and parking issues with workforces which have a high ethnic membership. Precisely at the point that access to ethnic groups is needed because of policy changes, stakeholders involved in those policy changes, either as promoters, or resistors, or as complainants, have reason to withhold access. It is another dimension which contributes towards the hidden in transport and travel. In a parallel fashion, a busy policy environment can create overload for consulted ethnic groups. The product of much consultation with ethnic groups is itself hidden: ethnic groups give interviews and are involved in consultations with small consultancy firms servicing local authority information collection needs. Partnership arrangements around regeneration have intensified this policy traffic. However, these research findings are rarely published, or provided to the ethnic communities in a form that they can use. The consequence is an industry in ethnic consultation which has no cumulative knowledge base in the public domain – the budgets are spent on ethnic consultation without the product being available for community use. In this context, attempts to obtain consultation with an Asian Women's Project were unsuccessful despite the Asian background of the researcher. However, persistence in the goal of revealing the hidden paid off and direct approaches to the Indian Community Centre resulted in a focus group which provided a wealth of voice and information on the ethnic travel and transport context.

'Hard to reach' groups are not only under-represented in consultation exercises (this 'hard to reach' attribute being, we argue, partly a consequence of past negative experiences) but are also under-represented in direct participation in regeneration programmes. Renewal.net (2002) give an indication of the level of under-representation of black and ethnic minority groups in direct participation through black and minority ethnic group led regeneration programmes:

- Out of 900 plus bids over the six rounds of the SRB, there were only 15 successful BME-led bids, representing 1.3% of the total
- The value of BME-led SRB programmes of £21m represented 0.4% of the total SRB programme budget
- SRB bids were concentrated in three regions – London, West Midlands and South East
- Most BME bids were relatively small ie less than £1m

(Renewal.net, 2002)

In the next two chapters on specific transport schemes in Bristol and Nottingham, we review the extent of ethnic consultation in the associated transport planning and give voice to ethnic concerns about operational issues connected with these schemes.

---

1 See also the work of the Los Angeles Bus Riders Union which seeks to promote environmentally sustainable public transportation for the entire population on the premise that affordable, efficient and environmentally sound mass transit is a human right (http://www.busridersunion.org/index.html).
2 It should also be noted that for some social groups (for example, some of the African Caribbean and young white respondents in our research) networks may be more important than households, indicating that it is not the case that one approach to travel behaviour fits all groups. This points to a need to develop appropriate approaches to take account of social structures such as the development of a Network Activity Travel Simulator (NATS).

Chapter 4

Road User Charging and Ethnicity: The Bristol Case Study

4.1   Introduction: outlining the Bristol road user charging scheme

I think we need better cycle and pedestrian facilities but I think it would be very sad if the only way they could get the money for this is by charging motorists and personally I'm against charging. All I ever hear from drivers in Bristol is 'If the public transport were different and would get me from A to B when I need to go from A to B, I would use it'. But they are saying it is so bad, they are not prepared to use it.
(Female resident, Easton, an ethnically mixed district of Bristol)

Radial routes mean all cars and buses have to go through the city centre causing congestion. If everyone gets in a car and wants to park in the city centre, this also causes congestion.
(Retired male resident, Southville, Bristol)

In the next two chapters on specific transport schemes in Bristol and Nottingham, we review the extent of ethnic consultation in the associated transport planning and give voice to ethnic concerns about operational issues connected with these schemes.

At the point the research for this book commenced, there was no operating road user charging scheme in Britain. The policy environment was characterized by much uncertainty about the public acceptability of road user charging. Local politicians and national politicians alike were sensitive to the chance of electoral disfavour and punishment which could follow upon the introduction of road user charging by any particular public body. Since this time, London and Durham have adopted congestion charging as a municipal practice and the national government has begun to talk of full societal road charging schemes operated on the basis of new information communication technologies such as geographical positioning systems. In this political environment, research was commissioned by the Department for Transport in discussion with the City of Bristol to identify issues of gender, ethnicity and lifecycle in respect of road user charging and to gain an understanding of the public awareness of road user charging policy. This chapter will provide an outline of the congestion charging scheme proposed for Bristol and describe the key findings of the empirical research carried out in the city on the potential impacts of the scheme.

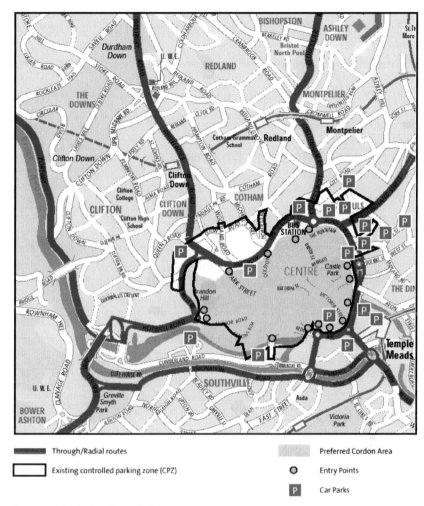

▬▬▬▬ Through/Radial routes	▨ Preferred Cordon Area
☐ Existing controlled parking zone (CPZ)	◎ Entry Points
	▣ Car Parks

Source: Bristol City Council (2000)
Reproduced by kind permission of Ordnance Survey. © Crown Copyright NC/03/22923

Figure 4.1 Proposed Road User Charging Scheme

The proposed Bristol Road User Charging Scheme consists of an inner city cordon with electronic tag/transponder technology for detecting vehicles as they cross the cordon and charging them accordingly. The scheme has fourteen entry points and includes the main shopping area at Broadmead, the Centre, West End and Harbourside but excludes the main inner city residential areas (Bristol City Council, 2000). The proposed charge would involve motorists paying £1 to enter the centre of the city between 0700-1100 on weekdays in the first year rising to £3-£5 in the fourth year.

Bristol City Council's Local Transport Plan (Bristol City Council, 2000) provides the following diagrammatic representation of the proposed Road User Charging scheme:

Few other details of the proposed Bristol road user charging scheme are available in the public domain. This is in part due, no doubt, to the political uncertainty surrounding road user charging as a demand management policy within Britain: an uncertainty which remains despite the apparently successful adoption of road user charging by London.

Accompanying the discussions around Bristol as a location for road user charging were discussions around the prospect of a tram or light railway (LRT). In marketing road user charging, tram or light railway proposals promise a more modern face to public transport. Investment in such infrastructure often carries a number of image benefits which provide the political space for other less glamorous transport measures to take effect.

4.2   The ethnic geography of Bristol: the context in which transport operates

Bristol is a major city located in the south west of England. The Audit Commission (2002) describes the city as diverse:

> …five wards are among the most deprived 10 per cent in the country, but Bristol overall is wealthy and has the second highest gross domestic product (GDP) per head in the region (on the latest figures available). Unemployment at 2.5 per cent is below the national average of 3.2 per cent but above that in the region as a whole (2.1 per cent). The city has a historic port and a history of achievement in engineering, commerce and the arts.

In the 2001 Census, Bristol had a population of 381,000, of which 51.2% were female, 33.7% were under 25 and 14.9% were 65 and over (National Statistics Online, 2002). The 1991 Census indicates that around 5.1% of Bristol's population was composed of ethnic minority groups: in comparison, 9% of the population of England were from ethnic minorities in the 2001 Census. The ethnic population of particular Bristol wards is substantially higher rising to 29% in the ward of Ashley. Between them the wards of Ashley, Lawrence Hill and Easton accounted for 43.9% of Bristol's ethnic minority population in 1991.

Table 4.1 Bristol – Ethnic Minority Profile (% of total city population)

Ward	White	Black - African Caribbean Other	Indian Pakistani Bangladeshi	Other Asian	Chinese	Other Ethnic Group
Ashley	71.0	18.9	6.2	0.9	0.3	2.7
Easton	78.5	10.0	9.3	0.4	0.4	1.4
Lawrence Hill	78.2	12.7	6.6	0.4	0.3	1.9
Eastville	89.8	4.1	4.4	0.6	0.4	0.8
Windmill Hill	93.3	1.9	3.6	0.1	0.3	0.8
Lockleaze	93.0	3.8	1.9	0.3	0.2	0.8
Bristol	**94.9**	**2.4**	**1.6**	**0.2**	**0.3**	**0.6**

NB Table only shows wards with highest ethnic minority population
Source: Census (1991)

In terms of the Indices of Deprivation 2000, the first of these three wards, Ashley was ranked at 756, the second, Lawrence Hill was ranked at 133 and the third, Easton, was ranked at 1043 respectively out of 8414 in England (where 1 is the most deprived and 8414 the least). With reference specifically to Bristol, the following table shows the deprivation scores for the 12 most deprived wards:

Table 4.2 Bristol – 12 Most Deprived Wards

Rank (against other Bristol wards)	Ward	Deprivation Score (2000)
1	Lawrence Hill	133
2	Filwood	221
3	Southmead	628
4	Knowle	733
5	Ashley	756
6	Whitchurch Park	921
7	Bishopsworth	935
8	Hartcliffe	1036
9	Easton	1043
10	Lockleaze	1095
11	Kingsweston	1207
12	Windmill Hill	1278

Source: Indices of Deprivation (2000)

Figure 4.1 gives an overview of the location of wards in Bristol.

BRISTOL WARD BOUNDARIES

Source: Bristol City Council
Reproduced by kind permission of Ordnance Survey. © Crown Copyright NC/03/22923

Figure 4.2   Bristol ward boundaries

As the objective of the research was to initiate an understanding of the impact of demand management on ethnic groups, a project never previously undertaken in the UK, it was important to recruit respondents in the areas of Bristol with higher levels of ethnic presence. Taking account of the evidence about the city's ethnic minority composition and deprivation, focus groups were held in Ashley ward and Easton ward with residents from Lawrence Hill attending focus groups in Ashley. The composition of the focus groups is given below:

Table 4.3   Bristol – Focus Group Locations and Target Groups

Location	Number of meetings	Comments	Main Target Groups
Ashley/Lawrence Hill	2	Held at Malcolm X Community Centre located in St Paul's area and serving residents of both Ashley and Lawrence Hill wards.	African-Caribbean community; parents using on-site day care facility, employed, women, public transport users, car drivers
Easton	3	Held at In Your Own Language, a community-based translation service	Indian, Bangladeshi, Pakistani, African-Caribbean & white Communities; users of translation services & students of computer training courses in adjacent building, unemployed, non-English speakers, public transport users, car drivers
Easton	1	Held at Easton Community Centre	Indian, Pakistani, Bangladeshi elderly, disabled, non-English speakers, car drivers
Easton	1	Held at Hairdressers' on Stapleton Road	African-Caribbean community, women

These wards were also important to the research because of their proximity to the city centre and M32 corridor and attendant challenges related to through traffic, commuter parking and rat-running on local roads.

The research on which this book is based was wider than the relationship between ethnicity, social exclusion and demand management measures: it also sought to assess the relationships between age, gender, social exclusion and demand management impacts. Interviews with focus groups other than those in ethnic areas also revealed information relevant to ethnic transport and social exclusion issues. For the sake of completeness these focus group details are presented here:

Table 4.4   Bristol – Composition of Other Focus Groups

Location	Number of meetings	Comments	Main Target Groups
Hartcliffe	1	Held at Hartcliffe Youth Project	Young, men, employed, unemployed
Windmill Hill	1	Held at Windmill Hill City Farm Senior Citizens Club	Older people, women
Southville	1	Held at Southville Community Centre	Older people

In earlier chapters, attention has been drawn to the dearth of materials on the travel behaviour and patterns of members of ethnic communities and in the preceding chapter we heard a number of ethnic voices drawing attention to the severity of problems they experience in the travel and transport environment, however, recruiting respondents to participate in empirical research and even in consultation on transport is no easy business. In many ways, ethnic groups can be characterized as 'hard to reach'. The difficulty is no doubt a combination of bad research practice in many cases and negative past experiences on the other hand, however, overcoming the difficulty is critical and extra effort must be made to reach those communities whose presence is registered and recorded through the census but whose voices are rarely heard in transport policy and strategy documents.

As has been shown in the previous section, recruitment focused on particular characteristics of groups and sought individuals who were representative of the wider target population. An over-riding characteristic such as age, ethnicity or gender tended to provide a key to finding that group, for example, Youth Workers/Clubs were used to contact young people, Lunch/Social Clubs to contact senior citizens. It became clear early on in the research process that recruitment to attend focus groups would be much more successful if people were already used to attending a particular venue at a particular time (e.g. always go to lunch at the Community Centre on a Monday lunchtime). In the cases where people had to be asked to attend at a particular meeting that was not one of their regular activities, focus group participation levels tended to be lower, for example, at the pilot at Southville where 12 people were invited to 2 sessions, only 2 people actually attended.

When people were asked to attend a meeting it was particularly helpful to have the support of a local person at the venue to help publicize it (in addition to posters/flyers), to take down names and to remind people of the meeting. At In Your Own Language, this technique was very successful. At the Malcolm X Community Centre, a similar approach was used. Unfortunately, on the day of the focus groups, several people who had agreed to attend were off sick with influenza and at the meeting that was arranged for people to attend after work, there was a very poor turnout of only 3 participants because there was a thunderstorm that evening in the Bristol area.

Different geographic areas in the city were selected to ensure a diversity of perspectives. However, it was very important to be willing to modify the methodology to increase participation in the study of those groups who were difficult to access by techniques that had been predetermined before the fieldwork started. In this way, the effects of unpredictable events on focus group attendance such as illness and bad weather were also overcome. It is important to recognize that geography is not simply about recording the presence of ethnic groups, it is also about recording their voice in relation to their spatial location.

4.3 The proposed scheme: operational issues

There is a potential in Road User Charging for development of boundary problems and displacement effects. The research indicates that there are existing problems with parking in the St Paul's area of Bristol. If a cordon charge is introduced in the city centre, one can assume that drivers will search in areas immediately outside of the cordon for parking, with St Paul's being the obvious destination, particularly since the M32 brings inbound traffic to this part of the city. One could therefore assume that the charge's introduction could result in a two-fold and linked displacement effect – the displacement of cars that would have been parked in the city centre to the St Paul's area and the consequent displacement of residents' opportunities to park in their local area from St Paul's to adjacent areas. This latter effect would be particularly pronounced during the morning school run period, when mothers may take their children to school in St Paul's, and as is characteristic of trip-chaining, then run an errand before returning home to find nowhere to park. This would exacerbate a difficulty already being experienced:

> The problem is because we live so close to town, everyone wants free parking. So if you're taking your kids to school, you'd better be quick otherwise you're stuck with your car in the middle of the road for the rest of the day.
> *(Young mother, at Malcolm X Centre Focus Group, St. Paul's)*

Figure 4.3   Parking on residential street, St. Paul's, Bristol

This displacement effect is likely to affect Easton as well. Taking the scenario further, if commuters are parking in Easton and St Paul's and then travelling into town, some are likely to use buses to access the city centre. Local residents of these two areas already describe being unable to board buses that are full by the time they arrive at their boarding point and this difficulty could be heightened, especially for those living on streets in close proximity to the city centre and nearing the end of the bus route. The resultant inaccessibility of bus services because of high loadings would be particularly detrimental to ethnic minorities, the young and the elderly who may not have an alternative but to wait for another bus with the likelihood that it will also be full. For others, the inability to use the bus may mean they have to resort to using taxis for essential journeys, biting into limited household budgets and evoking serious equity issues. In addition, this problem would be exacerbated if current practice of some bus drivers being unwilling to stop to pick up in these areas, whether due to late running or some other undetermined reason, as described by participants, is allowed to continue. These findings highlight the importance of hypothecation of the revenue generated by Road User Charging to improve local bus services. Possible solutions to these displacement effects are:

• using traffic warden real-time data to determine which drivers are committing parking offences in the St Paul's and Easton areas and ensuring adequate enforcement is complemented by strict residents parking zones

- provision of park and ride facility east of the area to divert commuters from parking on local streets
- incentives for car-sharing that allow access to the city centre for multiple occupancy vehicles at lower charge rates.

4.4 Public awareness and public acceptability: hidden perspectives

Although, the evidence from Bristol was that local residents had little awareness of Road User Charging and demand management transport policies in respect of Bristol, on having the logic of road user charging presented to them, many residents had interesting views as to how such policy should proceed and revenues be used:

> But to sell it to people, should spend money obtained from transport sources on public transport.
> *(Southville Resident)*

> (Road user charging) might (change my daily activities) but if LRTis introduced this is only on a fixed route and if I'm not in that area, it'll be of no benefit. It might be better to spend on other forms of transport, for example, small buses, tricycle, rickshaws, walking.
> *(Southville Resident)*

In addition to the focus groups and interviews conducted in Bristol around the proposed road charging scheme, travel diaries were also distributed (the details of the methodology are provided in Appendix 1 and the main travel diary findings are provided in Appendix 2). In these travel diaries, an open ended comment space was provided for respondents to express their views on any aspect of journeys they had made on the research day. Many reported difficulties in the trips they had made that day: the most common problems reported were those related to the quality of public transport and barriers to travel they had experience, the second most commonly reported issue in this open comment space was congestion. Public awareness of congestion is accompanied by an equal or greater awareness of the deficiencies of public transport.

**Table 4.5   Summary of Comments Made in Travel Diaries on
Difficulties in Trip-making in Bristol**
(listed by frequency of response)

Comment	Number of responses	Example of comment
Traffic congestion	11	On my second trip there was heavy congestion which delayed my journey
Parking problems	7	Parking was an issue in Broadmead.
Long wait time for bus	5	I had to wait for the bus for more than half an hour. That missed my half of class in Pakistani women organisation
Late running train	4	Trip 1: Local train 35 mins late Trip 2: Local train 4 mins late
No problems	3	No problems today
Unusually short waiting time for bus	2	On both my trips by bus the waiting time was very short (max 5 mins) which is quite unusual, especially in the evenings
High bus fares	2	I had to catch two buses – from Temple Meads to Broadmead and from Broadmead to Stapleton Rd. This whole journey is far too expensive
Buses unreliable	2	Didn't want to wait for bus – too unreliable especially in the evenings – so phoned a friend and asked for a lift
Unable to get bus/taxi with pram	1	Me and my friend did want to get a bus to town after going to Mothercare but we couldn't get on the bus with our prams and a taxi can't take our prams so we didn't end up getting to go
No bus information available	1	Could not find times of buses. No information on bus shelters. Lengthy waits at bus stop. Bought tickets as we travelled, did not know about day passes (until after sat on bus)
Buses not stopping	1	I am blind – one bus passed me at the stop, next bus almost passed but I managed to put my hand out at the last minute. On asking the driver why he wasn't going to stop, he told me that I should stick my hand out – no understanding that blind people are not likely to see the bus in time to stop it.

When asked to comment on the on the proposed Road User Charging scheme within the travel diary questionnaire, there were a mixture of responses: what was evident is that no blank cheque is being offered on the introduction of a Road User Charging Scheme. The public is not yet fully persuaded and there is a need for high quality information campaigns giving continuous updates and feedbacks on the progress and effectiveness of Road User Charging developments.

Table 4.6 Summary of Comments Made in Travel Diaries on Road User Charging in Bristol
(listed by frequency of response)

Comment	Number of responses	Example of comment
Would not affect me	12	My work is not in the city centre so I don't travel into the city centre on a daily basis. I don't drive a car so only journey I do make into the centre is by bus and the proposed charge is for cars so I presume bus fares will not rise. This cordon will not affect the types of trips I make
Need cheaper/better public transport to complement charge	7	I think a charge would be good, but also some good public transport as more people will need it
Will lessen congestion making buses more reliable	7	I would be more likely to take a bus, as this could make a significant difference to the amount of traffic congestion if many car users decided to leave their cars at home. As it is now, if you need to be some where on time, you are likely to be late.
Will not solve congestion problems	5	Charge may still not solve congestion problems. I don't think it will help to charge – people so want the convenience of cars that I think they will pay a high price before they think of alternatives
Agree with charge	5	I would like to see charges levied on private vehicles entering the city centre
Will result in less pollution	5	As I am not a frequent visitor to the centre the fee wouldn't greatly affect me although I do believe that the social benefits of a reduction in traffic and pollution would be extremely worthwhile as well as encouraging the use of public transport.
Needs exemptions	5	Also, there could perhaps be exemptions for people who really do need their car – like doctors for instance, or people who need to go from place to place.
Unfair	5	I think this charge would be unfair to drivers.
Acceptable if it contributes to general public welfare	3	Charges we do not mind as it is for the general welfare of the public

Will lessen congestion	2	The charge in the centre would not really affect journey I make as I tend to stay close to the area where I live. I do however have strong feelings about the benefits of such a charge and am of the opinion that such a toll would result in less traffic, less pollution and more frequent use of buses.
Another form of taxation	2	This is a very bad idea, as when we go for hospital appointments etc, it would cost anyone who was taking us extra costs, we also feel that after already paying road tax, why expect motorists to pay further charges. We strongly disagree with this idea
Willing to pay	2	I would still have to drive to and from work twice a week (0930-1445 shift) if a charge were introduced. I would be prepared to pay £1 to drive in at the above time. However if parking charges were introduced at a later date I may feel differently. I have minimal time in which to collect my daughter from school (1520) after finishing work (1445) and cannot walk to the centre and catch a bus in that time. I could take a bus to work if they ran at more definite times. If the bus in to work were reliable, I could alter my work times, finish early and catch a reliable bus back to my daughter's school. I would be more than happy to catch reliable buses to and from work, to avoid traffic queues and parking problems.
Could make parking difficult in my area	2	Parking outside my house could possibly become more difficult because if people used the street to park in to avoid paying a charge.
Annoying to have to pay	1	It would annoy me to pay a £1 every time
Would encourage me to go to Bath	1	On social events, I would probably go to Bath rather than Bristol.
Will alter journey time to avoid charge	1	If £1 charge is introduced, I will prefer not to travel between 0700-1100.
Will alter route to avoid charge	1	Change the root (sic) of my journey
Needs complementary policies eg. car-sharing	1	I feel that we have to do something to relieve the congestion. We need more park and ride places – if we do decide that cars must pay, could we not encourage 'sharing' by allowing full cars to travel freely?
Would stop me being able to get lifts	1	There'd be no more of that (lifts from her boyfriend). That would affect my reliability getting to work. The buses here are rubbish.

Responses in Bristol were largely uninformed about the workings of Road User Charging and initially this is likely to mean a higher level acceptance, however, as greater familiarity with schemes occurs and if no equity adjustments are made then this acceptance is likely to be eroded.

4.5   Conclusion: the importance of public consultation

In the light of the above discussion, it should be noted that the Audit Commission (2002) in an inspection report on traffic management at Bristol City Council, recommended that clear procedures for seeking views of the public for schemes and initiatives be introduced. The lack of interaction of Bristol City Council with the community on transport issues receives major comment in the Audit Commission Report – the Commission provides a list of areas in which improvements are needed:

- Consultation with the public needs to be more effective in influencing the delivery of traffic improvements.
- The public and other key external agencies find communications with the Council confusing, and this means that people do not always get an answer at the first point of contact.
- The priority given to new traffic schemes is not clear and consistent.
- New schemes to assist the management of traffic in Bristol have been delayed due to the shortage of staff, despite funding being available.
- Response to letters, phone calls and complaints is poor: they are not answered within the time scales set by the Council; challenging targets have not been set to improve performance; and there is no analysis of complaints in order to identify weaknesses and to use these as a means of driving improvements.
- There is a lack of an effective performance management system, and there are organisational, resource and programming issues that hinder efficiency.

From the perspective of public awareness of new transport schemes and the public acceptability of such schemes, the 'hidden' character of the relevant transport planning is far from a positive attribute: better public consultation is clearly required. Bickerstaff and Walker (2001), evaluating the experience of public participation in local transport planning in the United Kingdom, indicate that:

> For most authorities motivations for seeking public involvement have been grounded more in instrumentalism – satisfying the government's requirements for greater local consultation, increasing public support or 'improving' policy – than wider substantive and normative arguments about public competence, fairness, and democratic rights.

As we shall see later in Chapter 6, methodologies have been developed within the United States on Community Impact Assessment in the transportation area which greatly surpass the institutional structures observed in Bristol and described by Bickerstaff and Walker.

As we have seen in Bristol, there is a substantial ethnic presence which is largely unconsulted in the transportation policy and planning process. This represents a problem in an environment in which both ethnic communities and policy makers are becoming increasingly aware of the incidence of disadvantage attaching to ethnicity:

> Often the principle of community engagement has been translated into consultation processes, with decisions being taken elsewhere and leaving BME communities with a sense of having things done for them and not with them. With 70% of the BME population living in the most deprived areas of the country (compared with 40% of the population as a whole), we might expect that they would be centre stage for housing renewal and regeneration projects. The underlying logic of targeting regeneration interventions at community level in the expectation that they would inevitably have a positive impact on BME communities has failed – trickle down simply did not work.
> *(Renewal.net, 2002)*

Chapter 5

Work Place Parking Levy and Ethnicity: The Nottingham Case Study

5.1 Introduction: outlining the Nottingham Work Place Parking Levy scheme

Work Place Parking Levy schemes have not been widely used and consequently there are very few policy documents or transport analyses on the potential impact of such schemes. The discussions around the development of a Work Place Parking Levy scheme in Nottingham have been characterized by substantial resistance on the part of one of the major employers in the city, Boots:

> Nottingham's biggest employer, Boots, has warned the city council that proposed new charges for work place parking will jeopardise future investment in the city. Boots says it will challenge the parking levy if the authority presses ahead with it. The new workplace parking levy's due to come into effect in 2003. (Boots employs 7,500 locally.)
> *(BBC @ http://www.bbc.co.uk/ nottingham/news/2001_03/28/boots.shtml)*

At the time of this research, no Work Place Parking Levy had yet been introduced. This chapter will give an overview of the workplace parking scheme being considered in Nottingham and provide insights into its potential impacts based on fieldwork carried out in the city.

Workplace parking levy is aimed at commuters contributing to Nottingham's congestion and would involve a charge being made for each commuter car parked at the workplace. Employers would pay for a licence that reflects the number of commuter spaces they have. The employer would then choose whether or not to pass on the charge to employees. At present, the likely levy is £150 per commuter parking space per year (70p per day) rising to £350 per year in 10 years (Nottingham City Council, 2001).

5.2 The ethnic geography of Nottingham: the context in which transport operates

Nottingham had a population of 267,000 in 2001, of which 50.4% were female, 37.9% were 25 and under and 14.4% were 65 and over. According to the 1991 Census, almost 11% of the city's population was made up of ethnic minority groups, rising to 35% in one ward (Lenton). Unemployment, at 7 per cent, is nearly twice the national average. Of those employed almost 80 per cent are in the service sector with the hospitals and university providing 11 per cent of these jobs (Audit Commission, 2001).

Table 5.1   Nottingham – Ethnic Minority Profile (% of total city population)

Ward	White	Black - African Caribbean Other	Indian Pakistani Bangladeshi Other	Other Asian	Chinese	Other Ethnic Group
Lenton	65.0	14.5	16.3	1.1	0.7	2.6
Forest	70.7	7.0	19.6	0.4	0.3	2.1
Radford	77.1	8.9	11.3	0.3	0.5	2.0
St Ann's	77.5	13.7	5.9	0.4	0.9	1.6
Trent	80.3	4.1	13.5	0.4	0.2	1.5
Bridge	80.4	9.2	8.0	0.4	0.3	1.7
Nottingham	**89.2**	**4.6**	**4.5**	**0.3**	**0.3**	**1.0**

NB Table only shows wards with highest ethnic minority population
Source: Census (1991)

About 11% of Nottingham's total city population is from the minority ethnic communities. Some wards have very small minority ethnic communities e.g. Clifton West and East, Wilford and Bulwell East. The highest concentrations of people from ethnic minority backgrounds are in the wards of Lenton, Forest, Radford, St Anns, Trent and Bridge.

The Index of Multiple Deprivation 2000 was also examined to determine relative ranking of wards. All six wards with highest concentrations of ethnic minority populations are in the twelve most deprived wards in the city.

Table 5.2   Nottingham – 12 Most Deprived Wards

Rank (against other Nottingham wards)	Ward	Deprivation Score (2000)
1	Strelley	71
2	Manvers	88
3	Radford	163
4	Trent	211
5	Aspley	264
6	St Ann's	270
7	Bulwell West	395
8	Bridge	430

9	Bestwood Park	491
10	Lenton	527
11	Forest	542
12	Beechdale	571

Source: Indices of Deprivation (2000)

Figure 5.1 illustrates the ward boundaries in the city of Nottingham

Figure 5.1 Nottingham ward boundaries

With this background information, focus groups were arranged with Lenton Residents Association (Lenton Ward), Old Meadows Residents and Tenants Association (Bridge Ward), Embankment Residents Association (Bridge Ward) and users of the Indian Community Centre in Hyson Green (Radford Ward). The rationale for this choice of groups is given in the table below:

Table 5.3   Nottingham – Focus Group Locations and Target Groups

Location	Number of meetings	Comments	Target Groups
Lenton	1	Held at Lenton Community Centre	African-Caribbean community and white; residents living near to the City Centre
Bridge	2	Held at Portland Leisure Centre and Queens Walk Community Centre, The Meadows	Indian, Bangladeshi, Pakistani, African-Caribbean & white Communities
Radford	1	Held at Indian Community Centre, Hyson Green	Indian, Bangladeshi, Pakistani communities

In addition, given the research objective of obtaining gender and lifecycle perspectives as well as ethnic views, focus groups were also held as shown below with a young mother's group and an elderly women's club.

Table 5.4   Nottingham – Additional Focus Group Locations and Target Groups

Location	Number of meetings	Comments	Target Groups
Clifton	2	Held at Green Lane Community Centre	White, young women, elderly women

Despite an awareness of the importance of the various relationships between poverty and ethnicity in Nottingham, the main document summarizing these relationships does not discuss transport at all (*Poverty in Nottingham*, The Observatory, 2001). This is not surprising given that there is no source which can provide at present an evaluation of the share of public transport enjoyed by ethnic neighbourhoods on a systematic basis.

5.3   The proposed scheme: operational issues

New public transport schemes are understood within the history and the legacy of the old: improvement and decline are measured by the lived experience of transport by a city's residents. If planners and policy makers do not appreciate the impact of past transport history on perceptions of new schemes then they are unlikely to understand the nature of resistance to such schemes. Furthermore, poor public transport experiences are unlikely to generate trust in further changes (Ison and Rye, 2002). Extra dimensions of misunderstanding can be introduced by other policy agencies: whereas the research conducted in this study on Hyson Green revealed many public transport problems and challenges for local people, other research reports:

> The adjacent Nottingham inner-city neighbourhoods of Hyson Green and Forest Fields are areas of long-standing multiple disadvantage and have a negative image in the city. However, research reveals that local residents take considerable pride in their area. They point to many positive features, which could contribute to policies of area regeneration. The research team found that …convenience for the city centre and excellent public transport are much appreciated.
> *(Silburn et al., 1999)*

This text was produced in the context of a bid to get regeneration funding for the neighbourhood: it should be noted in such a policy context there is always present a danger of talking up the positive case in the bid to gain funds which are undoubtedly necessary but may come at the cost of a description of 'all's well' in certain areas such as transport, precluding future investment in these areas themselves. It is undoubtedly the case that full public transport audit of deprived neighbourhoods is required for the bulk of inner city locations for a true understanding of public transport deprivation to be assessed: as of yet no such audit has been undertaken and the perceptions of the residents interviewed by Silburn et al. (1999) are at variance with the voices recorded below.

Building a white elephant history: simultaneous negative change in other public transport arrangements

In many of the focus groups conducted in Nottingham it was clear that the issue of the alteration of bus routes with a reduction in the level of direct services and withdrawal of through bus services between the suburbs became connected with workplace parking levy and the tramway construction in the public mind.

> I'm sure the congestion now is temporary and it will be better after. A change to public transport happened 2 years ago. Before there were bus routes starting from one suburb to another, now they decided to make all people change in the city centre and some elderly people have difficult walking from

one end of city to other to change bus when before they just sat on a direct route linking one part of city to another. If you listen to these elderly people, people find standing at bus stops hard and even so long after changes, they say old system was better. And, in addition to that, they have to pay more if they have to travel at peak times. This is a major concern of people and has been in local newspapers etc.
(Asian man in early 40s, Hyson Green, Nottingham)

The reduction in the supply of public transport between suburbs without interchange is unlikely to play positively into a demand management scenario.

Some disabled people are no longer able to travel like they used to from one end of town to the other
(Asian Man, Hyson Green, Nottingham)

Removing through buses has an immediate impact upon the disabled who experience considerable difficulties with interchanges, most particularly where those interchanges depend on substantial mobility and agility between the stops involved in the different segments of the journey. Vulnerable road user audits should be used in assessing and evaluating policy decisions to discontinue through routes but rarely are. Negative changes such as these provide a public trust environment in which new measures are regarded with suspicion.

Severance and disruption during construction: a negative history

Residents reported high levels of disruption to traffic, alterations in public transport routes and in the location of bus stops and severe difficulties with normal parking arrangements during the construction of the tramway associated with the introduction of work place parking levy – the work place parking levy being viewed as the source of revenue for the operation of the tram. The alterations in public transport routes and in the location of bus stops due to construction work resulted in high levels of uncertainty in the transport environment and a loss of trust in the frequently changing messages received from the 'Corporation' on the timetable for the return to normal.

At the moment the changes are confusion. Suppose I was coming by car here, I come daily but I stop coming by car (driving took 10 or 15 mins) because it was so congested, so narrow and roads are dug up and we don't know where to go now. I come by ambulance now. But by force. The centre is booking it for me. I suffer from clots and travelling by ambulance takes 1 and a quarter hours to reach. All this time I'm sitting and going to many houses, I was telling my wife that this time sitting can form new clots. This not fault of ambulance driver.
(Elderly Asian man, Hyson Green, Nottingham)

The lengthening of journey times and journey routes as a consequence of disruption has some rarely noted but readily comprehensible negative impacts: the quotation above gives us some idea of the negative health impacts experienced by one Asian gentleman traveling daily to the Indian Community Centre in Nottingham.

Accessibility and the tram: a poorer service?

Many of the residents interviewed or consulted in focus groups commented on the lesser accessibility of fixed line tram public transport provision compared with the flexibility of buses and cars which penetrate backstreet areas even once the construction works were completed:

> I think the NET (Nottingham Express Transit) would be a totally useless idea. You have to catch a bus to get to the tram for starters.
> *(Young white mother, Clifton, Nottingham)*

> I'm also concerned because a rigid system like this one will need shuttle services to link people up to it and it could cause gentrification in the deprived areas it goes through, making it even more difficult for the people living in those areas to buy a home, one of the important steps to get them out of deprivation.
> *(Middle-aged white man in ethnically mixed neighbourhood, Lenton, Nottingham)*

> But my worry is that when the tram started, how people from backstreets get on? They have difficulty to get to it.
> *(Elderly Asian woman, Hyson Green, Nottingham)*

For some, the disruption was viewed as perhaps a temporary phenomenon:

> It was easy to traffic for car before, now people are bit confused as well. But my worry is that when the tram started, how people from back streets get on? They have difficulty to get to it. So second thing is then that they still using car and bus but bus going around around the wrong way in Nottingham and people confusing even more. But when they finish it I think it will be alright as well.
> *(Elderly Asian female driver, Hyson Green, Nottingham)*

Losing resources for the public good?: tram alignments in previously residential space

The equity considerations surrounding the resolving of central city congestion by utilizing residential streets for new transport corridors are highlighted by Nottingham residents:

> You know this traffic problem, OK for people going to city centre but people in residential areas are finding the difficulty. You know on streets with

terraced houses and people have the car, now I see the tram lines going through, where can people park now? Like on Knowlsey Drive, people worried what they going to do?
(Elderly Asian woman, Hyson Green, Nottingham)

This aspect of inequity within demand management schemes has received very little attention. More attention is paid to the equity issues involved in charging rather than to the equity issues of public space displacement.

Poor public transport, 'captive drivers': evidence from the trade unions.

At Nottingham Trent University, UNISON, the trade union conducted a survey on car parking and public transport issues in the context of work place parking levy proposed for Nottingham. The survey revealed workers' perceptions of a poor public transport environment resulting in a 'forced' use of the private car for travel to work purposes. The responses reported were:

CAR PARKING AND PUBLIC TRANSPORT ISSUES
Mode of travel
76% use a car at some time to get to work.
74% would use another form of transport
69% blamed their need to use a car to travel on the lack of adequate public transport.
38% would not arrive at work on time because of childcare or dependant responsibilities.
2% had equipment/loads to carry
Reserved Spaces
5% had a reserved space
Ability to park
Only staff who could arrive before
7.30am were able to park.
28% of employees thought flexible working would help to solve the issue.
10% would welcome more nursery places.
(UNISON, 2001: Information @
http://science.ntu.ac.uk/unison/Flyers/2001/11.
November%202001/Flyer%2001.11.01.pdf)

In undertaking this car use audit of members, UNISON was attempting to gain support for Nottingham Trent University to join forces with other large companies (Capital One, Boots, Nottingham University) who were protesting against the City Council proposal to introduce a Work Place Parking Levy. Nottingham Trent University had in the course of this negotiation with the UNISON acknowledged that work place parking levy charges would be passed on directly to staff.

An unsafe public environment: evidence from other studies

Personal safety and security emerges very rapidly as a constraint on demand management policies in a context where public transport is regarded as unsafe and walking regarding as a dangerous travel option:

> In Nottingham as many as 54 per cent of women interviewed on two low income housing estates said they never used buses after dark, and a further 30 per cent only used them rarely. Seventy-eight per cent reported feeling 'not at all safe' or 'not very safe' waiting at bus stops, and almost as many felt as insecure walking to the bus stop (Hiley, 1995). Women take a variety of measures to avoid situations seen as dangerous: they avoid walking alone at night, or only go out if a safe return has been arranged in advance; they avoid unsafe areas like subways and back streets and waiting at bus stops. Fear of crime increases the use of cars and the demand for close-by parking (Citizens Crime Commission, 1985).
> (Smyth, 2000 @ http://www.community-relations.org.uk/reports/transport/recommendations.htm)

Figure 5.2   Pedestrian route within The Meadows, Nottingham

Although Smyth (2000) does not report on ethnicity, it is clearly the case that the ethnic experience of an unsafe environment must be at least equivalent to that of women. Even where communities are consulted in relation to policy measures, ethnic groups are frequently underrepresented and this has been the case for Nottingham in respect of speed camera consultation (@ http://www.parliament.the-stationery-office. co.uk/pa/cm200102/cmselect/cmtlgr/557/557ap12.htm): in a survey of 511 members of the public conducted in November 2000, 468 (91.6%) were white, 19 (3.7%) were Black, 18 (3.5%) were Asian and 6 (1.2%) were from other ethnic groups. This means that the total percentage of ethnic minorities consulted was considerably less than would have been expected in a city with an ethnic minority population of about 11%.

In viewing the operational problems around work place parking levy from the perspectives of our ethnic respondents, the picture of the public transport system look very different to the 'excellent' evaluation given to it by Silburn et al. (1999), moreover, voices on Nottingham collected in the context of the Social Exclusion Units Report (2003) – Making the connections: transport and social exclusion – are more in fit with the portrayal of our respondents than with those of Silburn et al. (1999):

> **Case Study 3: Cath, unemployed, aged 25–34, single parent of two children (Bulwell, Nottingham)**
> Cath's children attend school daily, and travel by bus, which costs £1 each per day. However, she would prefer them to travel in other ways. "I would like them to walk or cycle but the youngest is too young to go on her own at present." She also feels that the buses do not run regularly enough, especially around school leaving time. "My eldest would like to do after-school netball, but there aren't any buses that come near our house at the time she would be leaving, so she can't do it."
> *Social Exclusion Unit (2003) @ http://www.socialexclusionunit.gov.uk/ publications/reports/pdfs/SEU-Transport_Main.pdf*

Additionally, there is a recognition by the City Council that there are major problems with public transport provision in Nottingham:

> Irregular timetabling of bus services is commonplace in Nottingham. Whilst a comprehensive level of service is provided across the network, the nature of time tables adopted by Nottingham City Transport is a major source of confusion for many people. Some services are operated on a part-day only basis, with very different networks provided in the peak and off-peak periods. Currently too many NCT services operate with uneven headways and non-clock face timetables during the evenings and peak periods.
> *(Nottingham City Council, Public Transport Plan, 2001 @ http://www.itsnottingham.info/ptplan6.htm)*

Moroever, the City Council recognizes that there are difficulties associated with radial routes and points out Hyson Green as a location that experiences such difficulties:

> Frequent bus links to Hyson Green and Sherwood are largely confined to the radial routes on which they lie.
> *(Nottingham City Council, Public Transport Plan, 2001 @ http://www.itsnottingham.info/ptplan6.htm)*

The policy climate has opened to admit of the actual experiences of public transport users: in this context ethnic voices should be increasingly heard.

5.4   Public awareness and public acceptability: hidden perspectives

Money-grabbing that's what it is.
(Asian man in his mid 50s, resident Hyson Green, Nottingham, commenting on speed cameras)

The issue of distrust of public agencies and progressive funding schemes was rapidly revealed in the focus group held at the Indian Community Centre. Public funding behaviour had already received a negative label in the context of funding decisions, and reversals around the community centre itself. The perception of speed cameras as a revenue generating device rather than as the safety measure it is advocated as is revealed in the quote above: the underrepresentation of such ethnic views on speed cameras in Nottingham has been already been detailed above. Arguing that there is an acceptability of a transport measure within the white community does not address the issue of the hidden lack of acceptability in the ethnic community.

Opening up the view into the hidden, the following discussion on the use of Work Place Parking Levy revenues and on the use of the revenues obtained from speed cameras is instructive:

(On the use of WPPL revenues in a meeting held at the Indian Community Centre, Hyson Green):

Asian male discussant 1: It is not a good way to make money.

Asian male discussant 2: When it is put into a big pot, the council will use it how they want.

Asian male discussant 1: I can give you an example, there was Section 11 money that we were supposed to get but the council used it in different ways. We were part of that Section 11 money but the funding stopped and we had to stop completely what we were doing.

Asian male discussant 2: That money was for community projects but we had to stop when no more money was given for it.

Asian Female – joining in discussion from far side of the room: They don't use money for what they said they will.

Asian male discussant 3: The police have these cameras everywhere now. Almost everybody has been booked, where is all that money going?

The interaction of negative public funding experiences with experiences derived from the history of other transport project measures and use of revenues has produced a poisoned ground in which work place parking levies are being evaluated and assessed by this community. Lack of transparency in public funding processes has a consequence.

In discussing hidden perspectives in Nottingham in relation to parking enforcement measures, it is important to consider the role that Asian social networks play as an alternative to public transport:

> I'm retired so I give the service for people. I take people to hospital or wherever they want to go.
> *(Elderly Asian male car driver, Hyson Green, Nottingham)*

> This is my feeling. Sometimes I take money from them (the people I give lifts to) depends. Sometimes I say I do for one hour, two hour your trip wherever you want to go, hospital or wherever but you give me £5. But if you want to do favourite for family or friend or other people who disabled and they ask, please can you do for me?, then I do free and no charge money. Also I recommend this lady can do this or this one. I use my network to do thing…(pause) …You know networks or communities they not built in one day, they take time..
> *(Elderly Asian female car driver, Hyson Green, Nottingham)*

> I think Asian families do that, they have very close knit families and bigger networks.
> *(Middle aged Asian male, Hyson Green, Nottingham)*

The Asian car drivers reporting above are elderly and retired. Work Place Parking Levy may not appear to be an issue for them at all, however, the description they give of ethnic support networks and the use that is made of these in accomplishing travel where public transport does not provide adequate services suggests that they may very well become the informal 'taxi' service for ethnic workers no longer able to pay to park in the workplace. The consequence of this would be that four journeys would be made to/from the work place per day in peak times in order to deliver one worker instead of the previous two made by the worker-driver. This impact is hidden from policy view if ethnic travel patterns are not properly studied.

5.5   Conclusion: the importance of public consultation

> The most popular approach to community engagement has been consultation but the act of consulting someone does not has automatically transmit a sense of engagement – especially where there is a history of consultation that does not result in desirable change and sustainable impact.
> *(Renewal.net, 2002)*

As an empirical measure of democracy, representativeness has three dimensions: geographic, demographic and political. The first requires that all territorial areas of a community are given the opportunity to participate. The second requires that no socio-economic group is excluded or disadvantaged. The third requires that all political views and preferences are afforded the opportunity of expression. While all three are important, however, it is the

demographic dimension which poses the greatest problems. Many studies of political participation have acknowledged the difficulties associated with obtaining the involvement of particular ethnic, socio-economic and age-related groups. Representativeness requires more than opportunities to participate. It requires more, even, than ensuring that individuals from all groups take part. There is a danger that those who come forward are self-serving individuals who do not necessarily share the values of the group whom they claim to represent (indeed, those who claim to represent groups which are traditionally difficult to involve are, by definition, different from those groups). There is also a danger that ill-conceived participation initiatives may favour some socio-economic groups over others, especially where they expect particular skills in participants. The socio-economic groups which are most disadvantaged are often those which lack even basic political skills and are least able to articulate their needs. This suggests that participation initiatives must go beyond involving those most willing to take part, in order to seek the views of those most disengaged from political processes.
(Pratchett, 1999)

Participation methods abound, but the question remains: so what? Does the mushrooming of initiatives actually broaden the base of decisionmaking? Activity and effectiveness are not the same thing. Indeed, we could be (and frequently are) swamped with initiatives, but the policy impact for specific sectors of the population might remain minimal or non-existent. The issue of social exclusion is particularly important in the participation debate. Failure to involve people from certain citizen groups is widely acknowledged. Patterns of social exclusion can be reproduced within participation initiatives—young people and citizens from minority ethnic groups are identified as particularly hard to reach. Many see participation initiatives as dominated by particular groups and definitely not for 'people like us'.
(Wilson, 1999)

Another dimension of the hidden is the 'talking past' which happens in community consultation that is merely token in its purpose. Respondents reported that often they were uncertain about the nature and the detail of consultation which had happened between public authorities and the community even within the community centre itself. Ensuring that communities are effectively consulted is very different from token consultation:

In a meeting at the Indian Community Centre, Hyson Green -

Researcher: So were you consulted on the tram?

Eldery Asian Male 1: I think consultation has been done.

Middle aged Asian male 1 (referring to the Nottingham Express Transit (NET) whose tram line was being constructed in the neighbourhood at the time): I have seen few times NET people book this centre.

Elderly Asian male 2: I happen not to know anything about the tram system that they are laying down.

Young Asian Female: It has actually been in the papers.

Middle aged Asian male 2 (outreach worker): It has been in the Evening Post, the Advertiser, they did load of things.

Elderly Asian Female: But can't everybody read Post, especially Asian people hardly read it. Then some people read it but can't bring it voice up.
(by 'voicing up' she meant 'have no channel for comment')

This discourse clearly reports on consultation which has gone wrong though the evidence in front of us is certainly that Nottingham made more attempt to consult with the community, and the ethnic community in particular, than the Audit Commission credits Bristol with having done.

Innovative approaches are needed to overcome current barriers to consultation with harder to reach groups. Indeed, as we shall see in the work of Needham (2002), transport itself can be a barrier to participation in consultation. Needham, reporting on consultation in Oxford, not only highlights problems councils face in expanding consultation 'but also shows how it can be used to entrench existing power relations':

> This demand problem (ie. lack of willingness to participate in consultation) is likely to be particularly intense among the 'hard to reach' groups that Oxford has sought to target. Officers' efforts to increase participation among young people, ethnic minorities and poorer communities have had only limited success. In part, this reflects an assumption that lack of participation indicates lack of interest, with officer efforts directed at mobilizing interest in consultation rather than tackling entrenched problems that limit participation. Childcare and transportation needs, for example, were not mentioned. The council publishes some public documents in languages other than English, but the selection of languages reflects earlier waves of immigration into the area rather than the current situation. Translation facilities are not offered at area committee meetings, perhaps recognizing that invitation letters sent out in English are unlikely to encourage attendance by non-English speakers.

The importance of public consultation becomes increasingly clear as the detail of travel and transport experiences is reviewed. Effective consultation requires that schemes pay attention to public suggestions for improvement and that consultation not simply be used as a one way street of providing information on that which has already been decided. Procedures and protocols for resolving conflicting interests and compensatory arrangements for remedying negative impacts will clearly be necessary but the development of such equity instruments can no longer be delayed.

It is quite clear that some organisations are simply going through the motions of public participation in response to the exhortations and veiled threats of various government pronouncements on the topic. It is equally clear, however, that many organisations see recent innovations as an opportunity to address some of the shortcomings of existing democratic practice. These organisations are actively seeking to renew democracy through the imaginative use of various techniques, at the heart of which is a simple belief that more participation must be better for democracy than less.
(Pratchett, 1999)

Chapter 6

Contexts and Audits: Local Configurations and Equity Techniques

6.1 Introduction: locality and specificity, a new transport vision

Transportation planners must consider both the benefits and burdens of their decisions.
(Centre for Urban Transportation Research @
http://www.ciatrans.net/index.shtml sponsored by the US Federal Highway
Administration)

US Policy Guidance Concerning Application of Title VI of the Civil Rights Act of 1964 to Metropolitan and Statewide Planning-

Assessing Title VI Capability_Federal Transit Administration (FTA)/FHWA Actions -Environmental Justice in State Planning and Research (SPR) and Unified Planning Work Programs (UPWPs): At a minimum, FHWA and FTA should review with States, MPOs, and transit operators how Title VI is addressed as part of their public involvement and plan development processes. Since there is likely to be the need for some upgrading of activity in this area, **a work element to assess and develop improved strategies for reaching minority and low- income groups through public involvement efforts and to begin developing or enhancing analytical capability for assessing impact distributions should be considered**
[emphasis added] in upcoming SPRs and UPWPs. Federal Register: May 19, 2000 65(98) @ http://www.epa.gov/fedrgstr/ EPA-IMPACT/2000/May/Day-19/i12590.htm

In the most deprived wards fifty percent of households do not have a car (compared to only 11 percent of households in the least deprived areas). Those who cannot afford their own private form of transport are dependent upon safe, accessible and reliable public transport to access key services.
(Social Exclusion Unit Report (2003) cited in Nacro @
http://www.nacro.org.uk/data/briefings/nacro-2003090400-csps.pdf)

Black and minority ethnic (BME) communities are more likely than others to live in deprived neighbourhoods; be poor; be unemployed compared with white people with similar qualifications; suffer ill health; and live in over-crowded and unpopular housing. People from BME communities experience widespread racial harassment and racist crime and are over-represented throughout the criminal justice system. A number of reports have portrayed past regeneration initiatives as having had little impact upon BME

communities and cited among other issues their failure to target interventions directly to BME communities and their failure to engage BME communities. The National Strategy for Neighbourhood Renewal places great emphasis on community involvement and has broken new ground in producing Race Equality Guidance for New Deal for Communities. Throughout the Neighbourhood Renewal programme, partnerships are expected to achieve meaningful engagement of BME communities, ensuring that they are able to contribute effectively and to secure sustainable benefit.
(Black and Minority Ethnic Community Networks, renewal.net, 2002)

The new range of transport planning tools found in innovations such as GPS and GIS protocols and approaches open up the prospect of more directed and targeted transport practice and organization. There are signs in the policy environment of such an opening up with the development of New Centres for Excellence in Integrated Transport Planning based in the local authorities as opposed to the universities: however, there appears to be no specific coverage of ethnic and equity issues within this framework:

In 2002, the DTLR set up a series of **Centres of Excellence for integrated Transport Planning,** designed to highlight good practice in integrated local transport planning and encouraging the sharing of experience. Some of these Centres of Excellence are of particular relevance in tackling poor transport links in neighbourhoods. For example:

Topic	Centre of Excellence	Details
Accessibility	Tyne & Wear	Improving access to public transport
Air quality	Bristol	Developing an Air Quality Action Plan
Bus	Oxfordshire	The provision of bus based solutions, including quality partnerships
Home Zones	Greater Manchester	Development and implementation
Integrated transport	West Yorkshire	Interchanges, smartcard ticketing, guided busways
Public transport	South Yorkshire	Information provision and integrated ticketing
Social inclusion	Merseyside	Ensuring quality in travel opportunities
Working through partnerships	Hampshire	Joint working, partnerships and innovative delivery techniques

Extract from Renewal.net on Poor Transport Links (2002)

Given the expansion of policy space for the evaluation of poor transport links and their consequences for deprived communities, this chapter will concentrate on the need for consideration of local circumstances and experiences, in order to address equity concerns, when developing transport policies. Listening to the voices in this book, there is clearly a gap between public and professional perceptions of accessibility and in perceptions of what measures are required to remedy deficiencies. The accessibility problems of one area are not necessarily shared by another: what constitutes a transport solution for one neighbourhood may result in an adverse impact in another neighbourhood. As we saw, the plan for the Nottingham tram was viewed as producing benefits for many but there were those who its planned alignment injured – smooth running traffic to the city centre was to be obtained at the price of lost street space in ethnic neighbourhoods.

The focus group evidence obtained in Bristol and Nottingham, and presented in earlier chapters, indicates that there is a clear need for local authorities to consider the wider impacts of proposed road user charging schemes and work place parking levies on different social groups. The capability of new information communication technologies to precisely map local context and measure social deficits creates a new auditing space within policy itself. A tool such as an equity audit would facilitate the examination of the key issues that need to be addressed in introducing schemes such as those presented in earlier chapters. Such audits would consist of checklists which should be carried out at various stages in any scheme's lifecycle. The audit should take account of all groups in society (such as those defined by gender, ethnicity, age, disability, income etc.). Furthermore, public transport auditing in and of itself is important given the extent of dependence of households (50%) in the most deprived wards in the United Kingdom on public transport.

6.2   Equity audits: expanding the existing transport audit family

As we noted in the introduction to this book, auditing has become part of transport planning and practice over a range of operational areas: road safety audits, cycling audits; pedestrian and vulnerable road user audits, gender audits, accessibility audits, and institutional audits, all figure in current transport policy and practice. Despite the panoply of measures and tools available equity audits in respect of changes in transport provision are far from common practice in the United Kingdom's transportation environment. In order to remedy this existing deficiency, this research has developed an equity audit for research in situations of transport change.

In order to ensure that the needs identified by ethnic groups involved in this research are explored and amelioration of their difficulties prioritized, an equity audit should be employed for all charging schemes. As described in earlier work (Rajé et al., 2002b), traditional measures of access indicate that inner-city wards to the east of Bristol are amongst the least deprived in terms of accessibility to key services. Local residents have reported and described a different experience. An equity audit should be carried out to explore their true levels of access to services and to quantify accessibility by ethnicity, gender, income and lifecycle experience. This would assist not only in acquiring a baseline understanding of accessibility but also in measuring how Road User Charging affects these groups and in determining how revenue obtained may be applied to close the gap between the public and professional perceptions of accessibility.

It is recommended that this audit should consist of a checklist that includes the issues in Tables 6.1-6.4. It should be noted that this example is provided as a first step in the development of a generic audit. It is informed by other audits undertaken elsewhere although the timeline described is largely based on the road safety audit process. The draft equity audit checklist provided here should be seen as a basis for further consultation and refining. The checklist is drawn up for a Road User Charging Scheme and could be adapted for application to a Workplace Parking Levy scenario.

The audit should be carried out at various stages in the scheme's lifecycle: preliminary design, detailed design, pre-opening, 6 to 12 months after inception and, consequently, at periodic 1-5 yearly intervals. The checklist would need to be adjusted to reflect changes in circumstances over time. The audit must take account of all vulnerable groups in society (gender, ethnic, young, elderly, disabled, unemployed, low income earners) – a step which is often not taken as we have seen in the case of the London Congestion Charging Scheme. We recommend that equity audits be carried out for all Road User Charging schemes. For equity auditing to be useful it cannot be confined to the start up phase of any scheme but is a tool for continuing and continuous evaluation and monitoring. In this way, not only would a baseline picture of issues be obtained but also progress towards equity would be measurable.

Table 6.1 Equity Audit Checklist - Residents Within Cordon

1	How does the policy directly affect people within the area of operation of the charge? (Look at issues such as charge exemptions, road traffic conditions, public transport improvements, urban environment improvements etc)	
2	How will traffic conditions change in the city centre? (Look at issues such as congestion levels, parking, ambiance for walking/cycling, public transport, use of bus lanes etc)	
3	How do people access city centre facilities (before policy introduction)?	
4a	How will people access city centre facilities (after policy introduction)?	
4b	Will this change how people participate in key activities?	
5	Identify groups most likely to be affected and describe how any difficulties may be addressed.	

Table 6.2 Equity Audit Checklist - Residents Immediately Outside Cordon

1	How does the policy directly affect people immediately outside the cordon (up to 2 miles away)?	
2	How will traffic conditions change in this area? (Look at issues such as congestion levels, parking, ambiance for walking/cycling, public transport, use of bus lanes etc)	
3	How do people access city centre facilities from this area (before policy introduction)?	
4a	How will people access city centre facilities from this area (after policy introduction)?	
4b	Will this change how people participate in key activities?	
5a	How do people access local facilities?	
5b	Will this change after policy introduction?	
5c	Will this change levels of service available locally?	
5d	Will this change how people participate in key activities?	
6	Identify groups most likely to be affected and describe how any difficulties may be addressed.	

Table 6.3 Equity Audit Checklist - Residents More Than 2 Miles from Cordon

1	How does the policy directly affect people within the local authority area but further away (more than 2 miles away)?	
2	How will traffic conditions change in this area? (Look at issues such as congestion levels, parking, ambiance for walking/cycling, public transport, use of bus lanes etc)	
3	How do people access city centre facilities from this area (before policy introduction)?	
4a	How will people access city centre facilities from this area (after policy introduction)?	
4b	Will this change how people participate in key activities?	
5a	How do people access local facilities?	
5b	Will this change after policy introduction?	
5c	Will this change levels of service available locally?	
5d	Will this change how people participate in key activities?	
6	Identify groups most likely to be affected and describe how any difficulties may be addressed.	

Provide a timed plan for implementation of measures to ameliorate any difficulties described in previous tables. For each key area that must be considered, a detailed and timed plan for action must be provided.

Table 6.4 Timed Plan for Amelioration of Impacts Determined by Audit

	Key area for consideration	Action required	Due Date
1	Consultation with local residents affected by the new policy		
2	Design of complementary measures		
3	Plans for feedback to/from residents		
4	Amendment of design to take account of residents' views		
5	Implementation of new measures		
6	Programme of monitoring of new measures and effects of charging policy		
7	Timing of rolling programme of equity auditing		

In later sections of this chapter, other existing and new audit tools for assessing the impacts of transportation decision making on ethnic communities, and for remedying the burdens imposed on communities by transportation change, will be discussed. For the moment, and in closing this section, we want to draw attention to the fact that this audit was presented to the United Kingdom Department for Transport as part of the final report of the Oxford University study on the Impact of Road User Charging and Work Place Parking Levies on Social Inclusion/Exclusion in respect of ethnicity, gender and lifecycle issues. As such it was accepted and has been published on the Transport Studies Unit University of Oxford Web Site with Department for Transport approval. It is, thus, now available and in the public domain in contrast to the many studies commissioned on community responses which neglect ethnicity and where they pay attention to it rarely enter the public domain.

6.3   The parking displacement audit: a new tool for the toolbox

Parking enforcement technology can be applied to audit impacts of charging on inner city communities. In terms of policy innovation, the research has identified a way in which displacement parking in ethnic areas can be used to audit the impact of RUC and WPPL on ethnic areas adjacent to Road User Charging schemes or where Workplace Parking Levy may cause commuters to search for parking away from the place of employment. Specifically, in areas adjacent to Road User Charging schemes or where Workplace Parking Levy, in the case of our research ethnic areas, where the burden of displacement parking falls heavily parking enforcement technology can be harnessed to measure the precise level and character (including geographic origin) of displaced parking and its impact on ethnic neighbourhood functioning. A precise measurement of displacement parking can be used to determine the compensatory revenue sharing allocation which is appropriate to the inequity imposed by the demand management scheme.

Figure 6.1   Lenton, Nottingham: an ethnically-mixed inner city area where displacement parking effects are likely to intensify under WPPL

In order to consider the practicality of this tool an appreciation of the State of the Art in Parking Enforcement technology is required. Firstly, the use of handheld parking enforcement equipment is becoming increasingly widespread. Basic equipment consists of a data-logger which allows electronic recording and printing of parking tickets. These devices can store information to be used by the enforcement officer and eliminate the need to convert the information for further use. Key benefits from the use of the equipment have been described by the City of Vancouver (City of Vancouver, 2000) as:

- improved ticket payment and compliance with parking regulations with ability to
- target repeat offenders with towing
- faster response to public inquiries
- better enforcement of the illegal use of resident permits in permit areas.
- improved legibility of parking violation notices
- reduced error rate in transcribing tickets
- reduced paper handling by both parking enforcement and by-law fines.

These benefits would be particularly relevant if the technology was deployed in the enforcement of parking in areas liable for spill-over effects associated with congestion charging and work place parking levy. In this way, the equipment is a tool for transmission of licence plate numbers and vehicle locations to the police, assisting them in targeting vehicles that are avoiding paying to enter the city centre by parking in inner city areas immediately outside the cordon.

Secondly, in the US, the process of issuing parking tickets has been taken one step further. Parking enforcement officers in New Jersey have instant access to the state's judiciary mainframe computer and wirelessly transmit tickets over the system in real-time. This allows officers to query the state computer and immediately access important driver and vehicle information (Symbol, 1997). With support from the DVLA, application of online access to vehicle licensing databases would make the process of parking enforcement in the ethnic minority areas adjacent to road user charging cordons even more expeditious.

Clearly, origins of displacement parkers or demand management 'escapees' can be identified within the realm of existing parking enforcement technology, the information can be used to audit demand management schemes and further used to produce higher levels of equity in the transport environment.

Compensatory revenue sharing is key to the attainment of higher levels of equity. In order to offset the indirect exclusionary effects of road user charging, and in proportion to the level of displacement parking and crowding out of local travellers by distance commuters on public transport services, resources can be drawn from the revenue earned through road user charging to provide intelligent demand responsive transport services to the ethnic community and other locals crowded off public transport facilities: this would have a strong social equity effect and gives an insight into innovatory reorganisation of transport to meet the social exclusion agenda without any substantial policy shift or incurred expense (Grieco, 2003a).

**Figure 6.2   Another image of Lenton, Nottingham: an ethnically-
mixed inner city area where WPPL displacement parking
is likely**

6.4   Ethnic audits: integrating social policy tools into transport

The concept of an ethnic transport audit developed in this book is built
upon a number of existing types of social audits already used in transport
and urban management: some of these are to be found in the United States
and some of these are to be found already in the United Kingdom. In
addition, ethnic auditing has begun to occur in a number of social policy
fields, most particularly around regeneration and renewal schemes. In this
context, renewal.net (2002) provide an overview of the current state of
affairs for Black and Minority Ethnic (BME) groups:

> BME communities are over-represented on almost all measures of social
> exclusion. A snapshot of recent research and case studies into BME
> communities in Britain shows –
>
> • More than half of African Caribbean and Africans and over a third of
>   South Asians live in districts with the highest rates of unemployment.
>   Only one in 20 live in an area of low unemployment compared to one in
>   five white people;
> • The proportions of people from different BME communities having a
>   household income of less than half the national average are 34% of

Chinese people, 40% of African Caribbean and Indian people and over 80% of Pakistani and Bangladeshi people. These figures compare to 28% for England and Wales as a whole.

- Africans, Pakistanis and Bangladeshis are two and a half more times likely than white people to have no earner in the family
- Pakistanis are twice as likely to be on housing and/or council tax benefit as the white population, whilst Bangladeshis are substantially more likely to be receiving these benefits
- Pakistani, Bangladeshi and African-Caribbean people are more likely to report suffering ill-health than white people
- Infant mortality is 100% higher for children of African Caribbean and Pakistani mothers compared to white mothers
- BME young people are more likely to be at risk of experiencing most of the problems of deprivation and social exclusion
- School exclusion rates for Black pupils are significantly higher than for others *(Renewal.net, 2002)*

As is immediately observable, transport despite its importance is not showing on this audit of BME disadvantage yet the indications from a host of small studies taken together are that there is a substantial issue to be addressed. The new planning forum Regeneration-Uk.com has no top heading entry under transport whatsoever: the impact of transport changes on community functioning has been largely handled as peripheral to the UK planning process in contrast to a more community targeted approach in the United States.

For our purposes in this volume, the Community Impact Assessment already utilized in the United States is, perhaps, the most immediately useful:

Community Impact Assessment, or CIA, is an iterative process of understanding potential impacts of proposed transportation activities on affected communities and their sub-populations throughout transportation decisionmaking.
(Centre for Urban Transportation Research @
http://www.ciatrans.net/index.shtml sponsored by the Federal Highway
Administration)

This approach focuses on the need to meet the access and mobility needs of all people and not simply the needs of the mainstream traveller:

The goal of the transportation professional is to help meet the access and mobility needs of all people through system planning; program and project planning, funding, development and implementation; and operation and maintenance. The community impact assessment (CIA) process shows transportation professionals how to reach this goal with community support. It encourages understanding community issues, concerns, wants, and needs, and taking them into consideration throughout transportation planning, program and project development, and program and planning implementation and

maintenance. A key tool in this process, throughout transportation decisionmaking is effective public involvement.
(Centre for Urban Transportation Research @
http://www.ciatrans.net/index.shtml sponsored by the Federal Highway
Administration)

This policy development clearly opens a door to fullscale ethnic audits: not surprisingly perhaps one of the key web sites describing and promoting this tool is located in Florida where issues of ethnicity are centre stage.

Community Impact Assessment (CIA) is an important part of transportation planning and project implementation. The inclusion of CIA allows for a community's concerns (mobility, safety, employment effects, relocation, isolation, etc.) to be addressed in transportation decisionmaking. Various laws, publications and events have impacted the development of CIA policies and measures. The history of CIA in Transportation began with the National Environmental Policy Act of 1969 (NEPA) and the process continues to develop.
(Centre for Urban Transportation Research @
http://www.ciatrans.net/index.shtml sponsored by the Federal Highway
Administration)

Importantly, consultation in such an Impact Assessment is not a one-off event but rather a process of planner and community interaction. The Community Impact Assessment literature available from the University of South Florida stresses that government agencies must work closely with communities in order to maintain or improve the quality of life and identifies the following activities to help achieve these goals:

- Use collaborative problem solving
- Promote openness and inclusiveness in transportation decisionmaking
- Keep public informed throughout transportation decisionmaking with periodic "status" updates, especially when active involvement is at an ebb
- Build working relationships with local agency staff and the public
- Use local contacts and community leaders to help identify and verify the likely community issues and concerns
- Establish a commitment compliance process that tracks commitments until successfully implemented
(Centre for Urban Transportation Research @
http://www.ciatrans.net/index.shtml sponsored by the Federal Highway
Administration)

The recent history of developing Community Impact Assessments in the United States, an approach which is now widespread, was built upon the recognition that transportation projects often had negative impacts for ethnic neighbourhoods which had been underrepresented in the planning process and that such communities were increasingly ready to take action in law to prevent developments about which they had not been consulted. The case study of Chinatown in Philadelphia is presented as an instructive example on the Community Impact Assessment web site:

Case Study 6.1 Chinatown, Philadelphia.

The story of the Vine Street Expressway involves several innovative approaches to highway design and community interaction, and is representative of the unique problems that can be encountered in the large, older, urban areas of the Northeast. The planned route of the Vine Street Expressway passed through several older, urban communities and attracted intense opposition from most of them. Among them was Philadelphia's Chinatown, a century-old community of ethnic Chinese already feeling threatened by the many construction projects underway nearby.

In the 1960s, Chinatown located south of Vine Street and just to the northeast of the Philadelphia CBD, was surrounded on three sides by urban renewal projects. In 1966, the community learned about the proposed Vine Street Expressway, which they felt, if undertaken as planned, would form a fourth and final barrier to the community.

Despite a lack of political power, the Chinatown community entered the fray in March 1966. Upon learning of the Vine Street Expressway plans through the newspapers, the leaders of Chinatown began to organize for the dispute that was to follow. This dispute would extant to 1983, when a Final Environmental Impact Statement (FEIS) filed by the Pennsylvania Department of Transportation (PennDOT) and Region 3 of the Federal Highway Administration (FHWA) contained a compromise plan for the expressway that met most of the needs of Chinatown. The result: a sensitively planned and aesthetically pleasing, below-grade, limited-access highway design that has helped not only to preserve but also to expand this energetic and cohesive community. Moreover, the Vine Street Expressway provides ample vehicular access to a successfully redeveloped downtown Philadelphia.
(Centre for Urban Transportation Research @
http://www.ciatrans.net/index.shtml sponsored by the Federal Highway
Administration)

The Community Impact Assessment approach also recognizes that discussion between all stakeholders is likely to provide a better and more rapid outcome to transport decisionmaking than are processes of under-representation which result in protracted public contestation.

In the United Kingdom, Crime and Disorder Audits (launched as a consequence of legislation in 1998) have become an importance source of information on the ethnic experience of personal insecurity in the wider travel and transport environment.

Hotspots already identified in local crime audits may include bus stops or train stations as sites of particular concern, or the movement of people in and out of the town centre on Saturday nights. Asking local transport operators

and the British Transport Police to provide information on the number and nature of incidents, as well as temporal trends (whether there is an increase in incidents at particular times of the day or night, on particular days or times of the year) will help to highlight particular issues that can be dealt with by the Crime and Disorder Reduction Plans.
(NACRO, 2003)

However, such audits represent an underestimation of the problems experienced in the travel and transport environment: NACRO reporting in 2003 argue that:

The majority of railway and bus passengers do not report harassment or crime, either because they do not think the incident is serious enough, or that transport staff will not treat the incident seriously. Many also fear that reporting a crime will lead to a delay in their journey and that the offender will not be caught in any case.
(NACRO, 2003)

6.5 Conclusion: predicting the emergence of localized audits

Reviewing the field of new audit techniques in social policy, it is clear that localized audits are an increasing practice. With the injunction in the United Kingdom for local authorities to develop Local Transport Plans and the moves towards developing local accessibility audits signaled by the recent Social Exclusion Unit report (2003) on transport and social exclusion, local audits in the transport domain are now clearly on the policy agenda.

As of yet, these have not been linked to demand management policy but as information is obtained from other audits such as the Crime and Disorder Audits, from local accessibility audits, from Local Transport Plan information bases and as an awareness of environmental justice policies developing in the United States increases so the pressure on demand management schemes to audit their impacts on equity are likely to mount. In the United States environmental justice has the weight of law behind it (policy guidance concerning application of Title VI of the Civil Rights Act of 1964 which is the basis for the US approach can be found in Appendix 3):

(US DOT 1996:11). The Executive Order (EO) 12898, Federal Actions to Address Environmental Justice in Minority Populations and Low-Income Populations, issued on February 11, 1994, required each Federal agency to develop an agencywide environmental justice strategy. The EO has as its main purpose the reinforcement of existing environmental and civil rights legislation to ensure that low-income and minority populations are not subject to disproportionately high and adverse environmental effects.
(Centre for Urban Transportation Research @
http://www.ciatrans.net/index.shtml sponsored by the Federal Highway
Administration)

In the United Kingdom, discretionary elements still feature more largely than do processes of enforcement or legal settlement in the settling of the balance between the rights of ethnic groups and the demands of the transportation system. The indications are, however, it is a system that will have to change.

Chapter 7

Conclusion: Finance, Funding and Fine Tuning Demand Management in the Context of Social Exclusion

7.1 Introduction: compensatory revenue sharing arrangements

The few authorities MORI has seen bucking the national trend and improving on highways and street scene issues are those where a choice has been made to increase spending on these services. But money isn't everything. Addressing a conference of Highways professionals recently it was striking to me how male, and "technical", even a little "ghettoized" the profession is. Roads and paving matter to most people – especially in rural areas where the problem is often seen as worst – but the profession needs to be shouting more loudly about the issue. Instead of being about technical standards it needs to join up its agenda with the wider group of people across government who are concerned about "liveability" (local environments being clean, green, well maintained and safe places) and quality of life in the broadest sense.

(Ben Page for MORI, September, 2003 @ http://www.mori.com/pubinfo/ bp_road-to-nowhere.shtml)

The public awareness and public acceptability of demand management measures are undoubtedly going to be of importance in the re-organisation of an effective British transport system. The research carried out in Bristol and Nottingham indicates strongly that it is important to examine a number of social and equity issues when developing congestion charging schemes. New approaches must be developed and old issues must be adequately addressed to heighten the likelihood of public acceptability of charging schemes. Inside of this policy space, the importance of compensatory revenue sharing arrangements around road user charging and work place parking levy emerges as a useful new direction. This chapter will highlight the need for taking account of the local context and the utilization of compensatory revenue based schemes for adversely impacted communities when developing congestion charging schemes.

7.2   Finance, funding and fine tuning: the toolbox

Discussions of remedying social exclusion in transport provision very rapidly hit the barrier of finance: where are the resources for funding improvements to come from in a public service structure where ownership has already been highly fragmented and privatized. New demand management measures can provide new resources within the public sector for use in the improvement of public transport services and the wider public transport environment. As we have already seen, ethnic minorities have transport needs which are presently either not met by the public transport system or are badly met by the public transport system. In this section, a range of tools for remedying adverse impacts of road user charging and work place parking levies are considered, these include:

- Hypothecation
- Equity Audits
- Exemptions and concessions
- Reorganization of public transport
- Demand responsive transport for essential journeys
- Enforcement of parking restrictions
- Introducing CCTV – lighting up the neighbourhood

Hypothecation:

Hypothecation has arisen as a solution in a context where it has been recognized that one of the major obstacles to implementation of an effective system of transport pricing is community resistance to charge for use of transport infrastructure when there is an expectation and history of free use:

> Community resistance will be reduced by greater transparency in pricing and greater transparency in allocating funds.
> *(Warren Centre for Advanced Engineering, 2002)*

In relation to road user charging, the UK Government has given a guarantee of revenue hypothecation that means that monies raised from congestion charging will be ear-marked for reinvestment in local transport initiatives. In light of the Oxford University research a key facilitator of social equity would be to improve good modal alternatives to the private car with hypothecated revenue being invested in making the public transport system less onerous to use. However, while investment in public transport will assist with equity, there must be an acceptance that for certain journeys and groups of people, car-based travel is the only alternative. For example, revenue could also be used to improve and pay for taxi journeys for those who cannot afford a car but need individual transport for medical, lifecycle or disability reasons. Demand responsive transport could assist in providing a solution to this problem.

Similarly on Work Place Parking levies the findings of the Oxford University research suggest that there is resentment of congestion charging and levies when they are seen as another component of the general taxing mechanism. In light of this study, social equity would be greatly facilitated if work place parking levy revenue is used to improve public transport service options and address obstacles to basic service access. It must be stressed that for certain journeys and groups of people, accessibility will continue to be car-based because of land-use planning trends that have affected availability of local services.

Equity audits:

In order to ensure that the needs identified by the groups of people involved in this study are explored and amelioration of their difficulties prioritised, a tool such as an equity audit could be employed for all Road User Charging schemes and Work Place Parking Levies. As indicated in the last chapter, this would be a checklist for local authorities, facilitating a survey of gender, ethnicity and income issues related to congestion charging. It would not be a one-off task at the scheme conception stage but a continuous process that allows iteration through a number of rounds over a scheme's life. In this way, not only would baseline issues be obtained but progress towards equity would be measurable with any adjustments needed to suppress rising inequities being captured expeditiously.

Exemptions and concessions:

It is recognized that the principle aim of congestion charging is to reduce the number of cars using the roads. Nevertheless, in London for example, not all drivers have to pay the central London congestion charge (@ https://www.cclondon.com/exemptions.shtml). Within the London scheme there is recognition that for certain categories of drivers and certain categories of vehicles and individual a range of exemptions and discounts may be appropriate, for example, there is a 90% discount for those living in the charged area. It is important in any charging scheme to take account of the need for exemptions and ensure that these are founded on equity considerations. For some people such as doctors and other essential workers who need to be readily mobile to carry out their employment responsibilities, exemptions should be considered. Other groups such as women and shift-workers may rely on a private car because of personal security, family responsibility or public transport unavailability reasons. They may not be able to alter their travel arrangements and trip patterns in response to Road User Charging and yet may fall into the low income categories who are already experiencing inequity in transport. For these people, exemptions would contribute to a fairer experience of transport.

It is important in any work place parking charging scheme to take account of the need for exemptions and ensure that these are founded on equity considerations. For groups such as women and shift-workers who may rely on a private car because of personal security, family responsibility or public transport unavailability reasons, exemptions should be considered. They may not be able to alter their travel arrangements and trip patterns in response to Workplace Parking Levy and yet may fall into the low income categories who are already experiencing inequity in transport. For these people, exemptions would contribute to a fairer experience of transport. It is equally important that guidance is given to employers, if they have decided to pass the charge on to employees, about exemptions. As expressed by participants, it would be unfair for people on higher salary grades to be exempt from the charge by having its payment included in their package of benefits while the lower paid workers have to pay to park regardless of their social circumstances and wider responsibilities. It is recognized, however, that although this was the view expressed by participants, there is no means by which policymakers can prevent companies applying the charge as they see fit. The legislation is such that employers are liable to pay the charge. Beyond that, it is up to the individual organizations to decide whether they bear the cost or pass it on to employees by whatever means they feel is appropriate. It may be that tighter regulation is required.

Reorganization of public transport:

Hypothecation allows deployment of revenue to resolve existing transport inequity. Revenues should not simply be used to buy more vehicles or upgrade existing infrastructure on the current patterning of provision but should be to adjust the pattern of transport provision to meet the needs of poorly serviced communities. Both in Bristol and in Nottingham respondents indicated that the present radial form of servicing had many deficiencies as a public service. In this regard, this research indicates that a major area that should be addressed is the reorganization of public transport services to allow journeys to be made that reflect community connections such as those that are found within ethnic minority communities or low income communities geographically separated by slum clearance and housing redevelopment. It should also be noted that the problems associated with having to take a one radial service into town to transfer to another to travel between adjacent communities may be exacerbated after introduction of cordon charging with the displacement effect previously described forcing local people off buses which are filled earlier in the route by commuters parking in the neighbourhood. This may be of particular importance for ethnic minorities: many respondents in this study talked of the problems that radial urban transport network patterns posed for community sociability – most particularly with regard to the discontinuation of through services. This emphasizes the need for road user charging and work place parking levy revenue to be invested in providing local, circumferential services that trace the social ties that exist in local communities.

Demand responsive transport for essential journeys:

Part of the reorganization of public transport under demand management regimes must be located in the development of demand responsive transport services. Furthermore, for the elderly, infirm and disabled or socially vulnerable or physically isolated, such as ethnic minorities or women, there must be flexibility in demand responsive services to enable journeys to be made easily. Without this flexibility at present, characterized by very short periods in which bookings can be made, the need to book two days in advance and the limitation to travel only during day time, several participants report that they are having to forego trips or use alternative resources such as relatives and friends for lifts or pay for taxis. Hypothecated revenue applied to improvements in such services to make them truly demand responsive, perhaps through investment in online scheduling and booking software and provision of taxi vouchers/services to supplement existing mini-bus based service, would contribute towards social equity and have an additional benefit of decreasing the number of private car trips that are being used as substitutes when demand responsive transport failure is experienced.

Within work place parking levy schemes, for the elderly, infirm and disabled workers – and as we have argued in this book, for female and ethnic minority workers, there must be flexibility in demand responsive services to enable journeys to be made easily where fixed route public transport services are not accessible or are not safe. Apart from wage earners and salaried employees, the issue of volunteer workers in the charity sector require attention: in many health authorities, volunteer drivers are an important source of health related transport and it is important that they have the necessary exemption from workplace parking levies especially as they are a source of demand responsive transport.

Within the Work Place Parking Levy scheme of Nottingham, the issue of providing demand responsive transport routes or feeder routes to the tram line and tram stops is a subject for further attention. For some participants, the idea of using the tram is attractive but there are concerns about their ability to access the service either because it is distant from their home or because they are elderly or disabled and therefore not able to get to a stop. Use of hypothecated Workplace Parking Levy revenue for provision of feeder mini-bus services to take passengers to tram boarding points would allow this group of people to use the new service. By making such a service demand responsive, an even greater contribution towards transport equity would be made.

Enforcement of parking restrictions:

In the last chapter, attention was drawn to the prospect of developing parking displacement audits which would identify vehicles parking in neighbourhoods adjacent to the congestion charging cordon or adjacent to neighbourhoods where workplace parking levy schemes were operational. Vehicles attempting to 'escape' charges arising from demand management policies can significantly disrupt the social and economic life of adjacent communities outside of the demand management zone. The development of appropriate parking policies to protect such vulnerable neighbourhoods as part of the demand management strategy requires consideration. Violation of parking restrictions in the adjacent neighbourhoods are under normal circumstances less likely to be policed and enforced. Enforcement of parking violations would be necessary to achieving equity and public acceptability of demand management measures in the neighbourhoods adjacent to demand management schemes.

It was also suggested in the last chapter that parking technology could be harnessed in identifying the level of infringement and in determining the part allocation of revenues earned from road user charging to compensate the adversely affected neighbourhood. Parking fines could also be directly harnessed to develop and provide demand responsive transport or improvements in fixed route public transport for such areas. This practice of compensatory revenue sharing as an equity tool has not previously been considered within the framework of demand management either in respect of road user charging or in respect of work place parking levy.

In several residential streets near to Nottingham city centre, residents have difficulty parking at present and participants anticipated a worsening of this situation if the levy is introduced and displaces people parking at city centre workplaces. In areas like Hyson Green where the tram now passes through residential streets, there has already been displacement of residents' parking. In light of existing and projected difficulties, it is important that parking restrictions are strictly enforced on non-residents' parking and, where not already in place, residents' parking zones be introduced.

Introducing CCTV – lighting up the neighbourhood:

In Nottingham, participants in areas where tram construction is underway expressed concern about relatives and friends who live on streets where the tramline has now been built. They reported that residents who used to park on street because they have no driveway or garage now have nowhere to park near home. It is inferred that for these residents, parking the car at a distance from their home raises concerns about personal security while walking to and from the vehicle, particularly in the dark, and car crime. In order to lessen the probability of personal attack or vehicle crime, investment of Workplace Parking Levy revenue in CCTV and improved lighting would be appropriate.

7.3   Public suggestions: input from local voices

Ethnic groups have a legacy of transport experience which is self-evidently beyond that of the mainstream population. This legacy includes previous experience of demand management – the Jamaican embargo on car importation for example – or the use of non-motorized transport such as bicycle rickshaws and motorized rickshaws as flexible transport forms which could provide solutions to low income residential neighbourhood transport deprivation. Shared taxis and the social network use of minibuses are forms of transport organization found in Africa and in Asia which have a lesser presence within mainstream British transport arrangements. Adopting more flexible transport forms may require engaging with vested interests in public transport service provision: areas and neighbourhoods which have seen a decline in service provision, and the statements we heard earlier in this volume from Nottingham City Council attest to this state of affairs, require alternative transport forms and it is important that legislation be revisited to enable this to take place. Current patterns of regulation and legislation are clearly not servicing low income and ethnic minority communities well (Social Exclusion Unit, 2003).

When considering the legacy dimension of different ethnic group's transport experience, it is also important to appreciate that this knowledge is constantly updated by return visits to the home country.

> International migrants are no longer relatively isolated from their countries of origin as might often have been the case in the past. Modern methods of communication such as air travel, satellite television, the telephone and the internet mean that members of migrant communities can maintain relatively intense relations with people and institutions in their country of origin. They can communicate regularly, even on a daily basis, if their resources permit. They may be able to travel frequently.
> *(Layton-Henry, 2002)*

> (A migration culture or tradition)…strengthens and renews family linkages and acts as an important conduit for the transmission and transference of values.
> *(Chamberlain in Goulbourne and Chamberlain (Eds.), 2001)*

Within certain sections of the public, questions are posed as to why the policy pressure is on reducing the use of the car and not upon reducing the supply of vehicles:

> But there are too many cars on the roads. It is only government can cut down on the production of cars.
> *(Elderly Asian man, Hyson Green, Nottingham)*

> It is the government that promotes new car building but they should introduce legislation to stop old cars being on the road. government wants industry to build new cars but they also want people to come off the roads.

The policies look like they conflict. The government don't know what they want.
(Black-Caribbean man, 35 years old, St. Ann's, Nottingham)

These quotations indicate that, for some, travel awareness appears to be linked to ethnic experience. The Jamaican man speaking in the second quote is of a similar age to the main author: we both grew up in Kingston in the 1970s when access to cars was strictly controlled through the Jamaican government's importation policies. When car importation rules were later relaxed, there was still a restriction on the maximum age of cars that could be brought into the island. It would appear that this man's experience of transport policy in Jamaica has definitely influenced his views on transport in the UK. The views of another Jamaican man in St. Ann's in Nottingham also underline how the legacy of ethnic experience at home can impact travel awareness and trip-making in Britain:

I only use the car but parking is bad in town and there is too much 'combustion' but I still use the car. I can park anywhere on the road here all day. I work in the shop so me no go nowhere in de day and at night I pay to park in town. I don't use bus just since I was a boy dis high.

For this man, driving a car is seen as the only means of transport he is willing to use.

Similarly, it is important to allow the voices from low income white communities to provide their information on experiences of public transport elsewhere through holidaying or visiting other locations either within Britain or abroad. The advent of cheap air travel expands the transport experience of those who historically were confined to local experience and local vision. It allows both the updating of migrant experience of the home country and the international mobility of the mainstream population:

Tim Jeans, managing director of MyTravelLite and previously with the airline that brought the US-inspired low-cost innovation to Europe 10 years ago in the form of Dublin-based Ryanair, said: 'It is not overstating it to say this is a revolution. In Ireland they speak of the "Ryanair generation", people whose work lives and leisure time have been transformed by the advent of cheap air travel. And there are legions of people in England, Scotland and Wales of an age and socioeconomic group who are now flitting backwards and forwards by plane in the same way they once hopped on a train or even a bus.'

Travel industry experts say it is the biggest transformation in international mobility for the ordinary British and Irish since package holidays were invented in the Seventies.

People who had never travelled abroad took off by the million on charter airlines, and tranquil Spanish fishing villages such as Benidorm were turned into concrete budget metropolises in less than two decades. The advent of

cheap, independent flying, often between regional airports, represents the twenty-first century new wave in travel.
(The Observer, 'Low-cost Brits take Europe by storm' 11 August 2002 @ http://observer.guardian.co.uk/uk_news/story/0,6903,772611,00.html)

7.4 A learning profession?: adjusted transport planning

Figure 7.1   An image of a residential street in the multi-ethnic neighbourhood of Easton, Bristol

> The Government is committed to making this country a successful multi-racial society where equality of opportunity is a reality for all.
> *(Home Office, 2002: 5)*

As we have seen in the course of this journey, there is a legacy of 'talking past' the excluded by the professionals (a situation recognized even in the office of the Prime Minister at No 10) and there is a legacy of talking of a past that was better in public transport terms amongst the socially excluded. There is a need to resolve the legacy of talking past, to move away from transport bureaucracy and toward integrated transport discourse and debate. Rather than seeking to understand transport, ethnicity and social exclusion, professional discourse has largely sought to deny its significance. Transport, demand management and social inclusion – the agenda is already set for debate. Inside the debate, the need for ethnic perspectives is not deniable.

Obtaining ethnic perspectives, as we have seen, is not an easy matter: the discussion of hard to reach groups has been present throughout our discussion: however, ensuring access to hard to reach groups is a matter or priority. In concluding this volume, it is important to remember that in order to study the potential impacts of new congestion charging policies, this research has had to examine peoples' experiences and views by putting individuals into categories such as ethnic grouping or gender. Other researchers have expressed some discomfort in categorizing people in this way and the potential effects this can have, for example in terms of racial classifications:

> The position of minority ethnic communities has often been portrayed as one of common disadvantage compared with the position facing the white majority. In this way existing research has tended to racialize the debate about minority needs, focusing on problems…as if they affect all black people equally…
> *(Blakemore, 2000)*

It must be noted that with a relatively small number of participants, all views expressed in this volume are individual but it is anticipated that they are representative of the perspectives of others living in their area and with similar socio-economic characteristics. However, as displayed by the findings and as a reflection of real-life, the experience in these groups cannot be discrete from that of others and there is a blurring of the distinctions between groups, with commonalities crossing the race, age and gender divides. In other words, transport equity considerations and efforts to introduce fairness to the individual's travel experience will, of necessity, bring a greater transport justice to the wider community regardless of gender, life stage or ethnicity.

Postscript

As this volume went to print, the DfT announced that more work needs to be done to improve ethnic minority and visible religious groups' experience of public transport. At the same time, the Department also made available an information pack to provide ideas to encourage more people from minority ethnic and faith communities to use public transport (available at http://www.dft.gov.uk/stellent/groups/dft_mobility/documents/page/dft_mo bility_025601.hcsp). This is a welcome step. It is hoped that, as a result of the DfT's interest in this area, a body of knowledge about ethnic experience of transport will be developed and that such information will be readily accessible to those who are working in the field. The increased awareness of ethnic perspectives must, however, be translated into changes that make a positive difference to the lived ethnic experience of transport.

Appendix 1

The Methodology

1.1 Introduction

The object of the research was to gain insight into the travel and transport behaviour of particular social groups – most particularly, 'hard to reach' groups, defined by gender, lifecycle stage or ethnicity, in relation to the introduction of demand management measures in the context of a growing understanding of the relationship between transport and social exclusion/inclusion. The study sought to examine the contribution that demand management measures can make to the reduction of the negative impacts of road based transport: Road User Charging and Workplace Parking Levy can potentially be used as instruments for social inclusion. The research also examined the potential negative impacts of these schemes if sufficient attention is not paid to gender, ethnicity and lifecycle issues: the lack of careful planning can result in such schemes having social exclusionary impacts.

The research sought to explore the range of impacts of different WPPL/RUC schemes and scenarios on the travel behaviour, time organisation and socio-economic activity of women, ethnic minorities, the young and the elderly. Another goal of the work was that, on the basis of the empirical research, practical guidance to local authorities would be provided regarding key elements to consider when introducing RUC/WPPL in relation to gender, ethnicity and lifecycle. The on-line toolkit (www.geocities.com/transport_and_society/roadusercharging.html) which has been developed was seen as one of the important research results.

In summary, by examining the possible impacts of demand management policies on vulnerable individuals and groups, the research sought to investigate the role of transport in processes that contribute to both social inclusion and exclusion. The overall methodology used for the primary research phase was one of focus groups and follow-up travel diaries, supplemented by face-to-face interviews. In the following sections, the methodology is described in further detail.

1.2 Methodology

This study began with a literature review phase during which an examination of existing databases on transport and social inclusion/

exclusion was also carried out. This provided the framework for the empirical research which followed.

1.2.1 Focus Groups

Focus groups were held in a number of geographic locations in the two cities. Most of the research was concentrated on the more deprived wards and/or those with highest ethnic minority composition. Recruitment to focus groups centred on particular sociodemographic characteristics to meet the research remit of examining gender, ethnicity and lifecycle issues. So, for example, lunch/social clubs were used to contact senior citizens, mother and toddler groups to contact young women. In addition, people were recruited through a local contact to attend a meeting where a formal pre-existing group did not exist, for example, users of a community translation service were invited to attend focus groups in their own language by one of the translation project workers, a receptionist at a community centre helped recruit participants who used the local crèche facilities to attend a focus group scheduled to be suitable for the times parents collected their children. Focus groups with ethnic minority participants were conducted in own language. All group discussions were recorded on mini-disc, participants were paid for attending, provided with refreshments and promised feedback on the overall study findings.

Murray and Davis (2001) have said that, in evaluating transport services, it is important to discuss how persons who are needy, or more in need than others, may be identified. The authors cited other research by Morris (1981) and Starrs and Perrins (1989) which identified several groups as having a transport need such as the elderly, the young, those who cannot drive a vehicle, the disabled, low income earners, women and those of an ethnic minority background. Given this context, this section describes the basis for choosing each of the groups represented in the focus groups and the construction of the meetings. The reasons that the attributes given below were felt to be significant within the framework of the transport experience, and therefore worthy of study in regard to congestion charging, are briefly outlined. The rationale for choosing different groups to participate varied in the two cities because of the differences in potential impacts of the two policies being considered. To capture these nuances, each city is therefore described in a separate section below.

Specific Groups – Bristol

Age

There is evidence to suggest that age affects mobility. The elderly and the young often have lower income and high rates of exclusion. For those who drive, young adult and elderly motorists tend to make short, low value of time journeys (e.g. leisure, shopping) and therefore may be particularly sensitive to charging. These two groups can have greater personal security concerns.

Gender
Women have on average lower income/high rates of exclusion. Female motorists make short, low value of time journeys (e.g. school escort, shopping). Women have higher rates of part time employment and can have greater personal security concerns. They may also be more reliant on private transport for certain journeys (e.g. night time shift work).

Ethnicity
People with ethnic minority backgrounds have on average lower income/high rates of exclusion. Ethnic minority communities are often concentrated in inner city areas where charges are most likely to be levied and individuals from certain ethnic minorities may also be key transport providers (e.g. private hire taxi industry). Ethnic minorities have greater personal security concerns and are more reliant on private car or community transport for certain journeys.

Disability
Accessible transport is vital for disabled people so that they have the opportunity to play a full part in society. In the context of this research, the needs of the disabled also take account of the needs of people with babies in pushchairs and those with heavy shopping. For this group of people, there can be a great reliance on car-based journeys for essential trips and, as a corollary, any policy such as congestion charging directly influencing car use will be significant for them.

Sheltered housing residents
Elderly people living in sheltered housing may not have family or other social support systems to rely on. Blakemore (2000) found evidence to suggest that amongst the African-Caribbean and Asian communities, families are not as supportive to older people as they were in the past. The author suggested that the strong demand for sheltered housing development by these ethnic groups indicates a search for extra-familial ways of meeting needs among minority older people. In light of this finding, it was considered important to understand how people from ethnic minority backgrounds who live in sheltered housing travel activities and what social networks they rely on.

Language
Large cities tend to have concentrations of residents from ethnic minorities or whose first language is not English. These groups may either be from the traditional minority ethnic groups in England such as people from the English-speaking Caribbean and people from India, Pakistan and Bangladesh who represent the older wave of migration to the UK and are

often second – or third – generation British. There are also new wave migrants who are resident in English cities such as asylum-seekers and refugees from places such as Eastern Europe and Africa. It was considered important to capture the views of both new wave migrants and the elderly non- English speaking people from the old wave ethnic minority communities in the study area to complement the views expressed by English speakers. For this reason, translation was provided at some focus groups.

Employment status

Often the unemployed are faced with having to pay for transport in order to look for work. The research sought to find out how transport may affect people's choice of workplace or whether they chose to work shifts, in order to determine whether Road User Charging would affect their employment choices and opportunities.

Public transport users

While Road User Charging is aimed at car drivers, it was recognized that the policy could have ancillary effects on people who use public transport. In addition, for those who use public transport but hope to become car drivers, it was important to gauge how the cordon charge may affect those aspirations.

Car drivers

The policy of Road User Charging is intended to affect the behaviour of this group of people. As shown above, members of the key groups being studied – women, young people, the elderly and ethnic minorities – demonstrate a degree of dependency on car-based travel for basic journeys. The views of car drivers on the policy were therefore vital.

Taxi drivers

People who do not have access to a car of their own, or a relative's, friend's or neighbour's and cannot or do not use public transport for whatever reason, often rely on taxis. The reaction of taxi drivers to the introduction of a charge to enter the city centre will have implications for this group's mobility, thus taxi drivers' views were sought to help develop a picture of the possible impacts for the taxi-dependent.

Users of city centre facilities

Since a cordon charge would affect people using facilities in the city centre, it was important to gain an insight into who uses such facilities and how they travel.

Specific Groups – Nottingham

Age

There is evidence to suggest that age affects mobility. The elderly and the young often have lower income and high rates of exclusion. For those who drive, young adult and elderly motorists tend to make short, low value of time journeys e.g. leisure, shopping and therefore may be particularly sensitive to charging. These two groups can have greater personal security concerns.

Gender

Women have on average lower income/high rates of exclusion. Female motorists make short, low value of time journeys e.g. school escort, shopping. Women have higher rates of part time employment and can have greater personal security concerns. They may also be more reliant on private transport for certain journeys, e.g. night time shift work.

Ethnicity

People with ethnic minority backgrounds have on average lower income/high rates of exclusion. Ethnic minority communities are often concentrated in inner city areas where charges are most likely to be levied and individuals from certain ethnic minorities may also be key transport providers e.g. private hire taxi industry. Ethnic minorities have greater personal security concerns and are more reliant on private car or community transport for certain journeys.

Disability

Accessible transport is vital for disabled people so that they have the opportunity to play a full part in society. In the context of this research, the needs of the disabled also take account of the needs of people with babies in pushchairs and those with heavy shopping. For this group of people, there can be a great reliance on car-based journeys for essential trips and, as a corollary, any policy such as Workplace Parking Levy directly influencing car use will be significant for them.

Employment status

Often the unemployed are faced with having to pay for transport in order to look for work. The research sought to find out how transport may affect people's choice of workplace or whether they chose to work shifts, in order to determine whether Workplace Parking Levy would affect their employment choices and opportunities.

Public transport users

While Workplace Parking Levy is aimed at car drivers, it was recognized that it could have ancillary effects on people who use public transport. In addition, for those who use public transport but hope to become car drivers, it was important to gauge how the charge may affect those aspirations.

Car drivers
The Workplace Parking Levy is intended to affect the behaviour of this group of people. As shown above, members of the key groups being studied – women, young people, the elderly and ethnic minorities – demonstrate a degree of dependency on car-based travel for basic journeys. The views of car drivers on the policy were therefore vital.

1.2.2 Travel diaries

This section describes the methodology used for the travel diary component of the research. However, some description of the focus group methodology is also included because of the dependence of the travel diary study on the successful recruitment of people at the focus group stage of the work.

The travel diary survey was carried out to build on the focus group research and was designed to obtain an insight into the travel patterns of residents of Bristol and Nottingham on a normal weekday. The questionnaire was designed to find out about all trips made on the research day. Information sought included time of travel, mode used, origin and destination and participants were asked to provide any comments they felt were important on the trips made that day. In addition, respondents were presented with a scenario that described the policy being considered in their city and then asked their views on such a policy and how this may affect the trips they had made that day. This question was open-ended to allow people the freedom to describe their views in as much depth as they felt appropriate. Respondents were then asked for their age, gender, ethnic group and postcode and, finally, for the names and addresses of three other people in their city who could then be also asked to complete the questionnaire.

As stated above, the main method of distribution of diaries was through focus groups. All participants at focus groups were asked at the end of the meeting if they would fill the diary on the following day and questionnaires were then distributed with Freepost envelopes along with payment for attendance at the focus group meeting. Participants were reminded that the questionnaire was anonymous and when completed would only be seen by the researcher. In addition, they were told that all information would be treated confidentially and used solely for transport research purposes. It should be noted that arrangements were made for anyone who could not read or write English to be provided with assistance in translation and completion of the questionnaire. In addition, diaries were completed for travel on weekdays only and avoiding Bank Holiday periods.

The 'snowball' contacts provided by respondents were sent a covering letter giving information about the study and explaining that their name had been given by a participant in a local focus group. The researcher's full contact details were also provided and a questionnaire and Freepost envelope were enclosed. Since the outgoing mail to 'snowball' contacts was unsolicited by the recipient, it was also important that the letter and questionnaire were enclosed in envelopes that were readily identifiable as coming from the Transport Studies Unit at Oxford University since this would lessen the likelihood that they would be dismissed as junk mail prior to being opened.

As the research progressed, it became evident that response rates to the travel diary survey were relatively low in Bristol. In order to increase the number of people asked to complete the diaries, questionnaires were also distributed to staff at NHS Walk In Centres and to members of youth clubs.

1.2.3 Interview diaries

In both communities, but particularly in Nottingham, there were problems of recruiting focus group members from the African-Caribbean community. This is a problem that has been experienced by a number of other researchers (Campbell and McLean 2002). As a result, a series of supplementary interviews were undertaken (19 in Bristol and 14 in Nottingham) by stopping people on ethnicity. Since the research focused on the destination end of the journey, these interviews were carried out at employment locations (such as hairdresser's, railway station), critical transport locations (railway station, bus stops) and medical destinations (clinics, chemists), other key service locations (Post Office, parades of shops eg Stapleton Road, Lower Ashley Road).

This interview diary technique consisted of stopping an individual on ethnicity grounds and ascertaining whether they lived in Bristol, if not the interview was discontinued. The interview asked the person to think of their previous day's travel and also asked about Road User Charging – whether they had heard about it and if so, where. In this way, it was possible to build a social picture of travel and thereby develop a greater understanding of main activities people studied were involved in on an average day and how they travelled to participate in these activities. In this way, this research technique was able to capture the social dimensions of transport. In order to obtain an insight into their travel patterns and explore the key factors that had been investigated in focus groups, after these people were asked about their previous day's travel, a diary was completed by the researcher as a record of their description.

This interview diary technique was extremely successful in that it allowed clarification of trip-making issues to be carried out by the interviewer and enabled a greater understanding of a person's views to be developed than was possible on an individual basis at focus groups or from respondent-completed travel diary where no further explanation of what has been written is possible. It enabled the researchers to gain an insight into social patterning of travel by exploring, as in the self-completed diaries, actual features of trips made such as number of trips, their origin and destination, mode used and any difficulties encountered. It should also be noted that the technique was particularly beneficial in that it introduced random participation to the research: people participating in focus groups were, to some extent, self-selecting but stopping people also bestowed an element of randomness to selection. The one drawback of the technique as a substitute for respondent-completed travel diaries was that it did not capture times of travel in detail, thus analysis of these interview diaries could not contribute to calculations such as average trip length. If people stopped to be interviewed did not have time to be questioned in detail about their previous day's travel, they were asked how they usually travel, whether they had heard about the policy proposed for their city and their views on this policy if introduced. These participants were asked to complete travel diary on the day after they were interviewed and given a Freepost envelope to return it to the researchers.

1.3   Comments on success of methodology

As previously stated, the research set out to investigate the gender, ethnicity and lifecycle issues associated with congestion charging. The combined methodological approach of using focus groups and travel diaries was designed to obtain insights into participants' experience of transport at present, their views on what could improve this experience and on the particular charging policy proposed in their city. It must be recognized that focus groups are not a statistical tool, they are indicative. In addition, the overall findings are based on a very small sample of individuals mainly drawn from particular social groups rather than the population as a whole. The decision to use a focus group/travel diary methodology was informed by the objective of reaching social categories not normally well represented by survey methodologies. For women, people with ethnic minority backgrounds, the young and the elderly, this combination of techniques was therefore appropriate. The emphasis of using qualitative methods (such as focus groups) is on accurately portraying or 'giving voice' to people's experience (Cancian, 1992). Qualitative methods of data collection can be sensitive to the unique personal experiences, perceptions, beliefs and meanings related to individuals (Sim, 1998). As a tool for 'giving voice' to the experience of the people participating in this study, the focus groups were extremely successful. For ethnic minority groups whose views may be hidden from mainstream discourse by language, the provision of own language focus groups was important to gain insight into people's perceptions and experiences of transport.

1.3.1 *The need for an evolving methodology*

One of the most important factors for successful collection of primary data for this study was the recognition that the methodology had to evolve over the course of the research in order to meet the key objectives. This evolution was on a number of levels which are described below.

At the start of the empirical research, there were difficulties in making contacts in the target communities in Bristol (whereas in Nottingham local authority support through existing the existing Area Committee network enabled easier connections to be made with local residents). Therefore, in Bristol, alternative methods of approaching local people needed to be employed. This relied on the support of a local person, often referred to in other literature as a 'gatekeeper', preferably associated with a venue, and was very important to the success of this part of the research. It was also necessary to select venues which were familiar to the study's target groups to help increase the likelihood of participation. Different geographic areas in the two cities were selected to ensure a diversity of perspectives. In addition, it was very important to be willing to modify the methodology to increase participation in the study of those groups who were difficult to access by techniques that had been predetermined before the fieldwork started: the evolution of the interview diary method when difficulties in accessing 'hard-to-reach' groups were encountered illustrates how techniques can evolve

1.4 Conclusion

The research carried out in Bristol and Nottingham used the complementary methods of focus groups and travel diaries supplemented by interviews. The methodology evolved over the research period to enable appropriate approaches which helped facilitate the inclusion of the more difficult to reach groups in the study. The combined use of qualitative and quantitative methods provided the researchers with insights that could not have been easily obtained if only one type of method had been used.

Appendix 2

Main Travel Diary Findings

2.1 Research findings – Bristol

2.1.1 Travel Diary Distribution and Return

The table below provides information on travel diary distribution and response rate:

Table A2.1 Travel Diary Distribution and Return – Bristol

Travel Diary Distribution and Receipts	Number
Number of travel diaries distributed @ focus groups	75
Number of travel diaries distributed @ NHS centres	20
Number of travel diaries distributed @ youth clubs	15
Number of travel diaries distributed by 'snowball'	71
Number of travel diaries distributed @ interviews	5
Total number of diaries distributed	**186**
Number of travel diaries returned	**56**
Rate of return	**30%**
Number of returned diaries providing 'snowball' contacts	30
% of respondents providing 1 or more 'snowball' contacts	54%

In addition, 14 diaries were filled by interviewing people at key destinations in the city, providing an overall number of completed travel diaries of 70.

At 30%, the rate of return of travel diaries was relatively low. This was despite the fact that over 60% of the diaries (115) were distributed through face-to-face contact.

There were several factors that may have contributed to this undesirable response rate:

- Although arrangements were made for translation of diaries to be provided for Somali and Sudanese participants, none returned the questionnaires. In comparison, members of the Indian and Pakistani communities did make use of translation services offered and subsequently returned diaries completed in this way.
- There was generally a low level of awareness of proposals for road user charging in the city and, after having the opportunity to talk directly to

131

the researcher at focus groups about other transport concerns, some people may not have felt sufficiently motivated to complete a questionnaire relating to a policy they felt was not relevant to them.

• It was difficult to get people invited to participate through the 'snowball' technique to fill in the diaries since there was no incentive offered for doing so.

The socio-demographic breakdown of respondents was as follows: 56% female, 7% aged 25 and under and 20% 66 and over. Just under half of all respondents were from ethnic minorities.

2.1.2 Average number of trips and distance travelled

All respondents were categorized by postcode into inner or outer to reflect whether they lived in a ward near to the city centre or further away. From Table A2.2, it can be seen that respondents in Bristol made an average of 3.18 trips person per day and travelled a mean daily distance of 19.16 km. Residents of inner city wards made fewer trips (3.00) and these made up a shorter total distance (16.96 km) than those made by outer ward residents (3.67 trips, 24.85 km). In other words, those living in outer wards were more mobile than those living in inner areas.

In terms of trips made, it is important to record here that talking to respondents in inner wards in the city indicated that some of them had great difficulty making certain trips to key destinations they wished to visit (eg. to visit friends/relatives in adjacent neighbourhoods, to hospital) because of issues such as community severance rendering walking difficult,[1] inconvenient bus routings and the expense of using buses and taxis. This was not reported by outer ward residents. With this background, it is difficult to determine whether the number of trips made in inner city areas is lower because key services are indeed more accessible (the straight-line accessibility measure on the Index of Deprivation puts some of these inner city wards in the top quintile in terms of accessibility) or whether this is an indicator that trips are not being made because it is difficult to do so.

Looking at the potential impact of road user charging, the evidence suggests that with outer ward residents making more trips and these being of longer length, depending on their destination and mode of travel, residents of these areas would experience higher travel costs if a charge is introduced. However, if the charge helps solve congestion problems, these residents would also benefit from increased journey times. Perhaps then, there needs to be an evaluation of outer residents' willingness to pay against journey time benefits gained.

In terms of time, individual trip lengths reported by respondents ranged from 1 minute to a maximum of 720 minutes. The average trip length was 46 minutes. It should be noted though that examination of the raw travel diary data indicated that there were varying levels of accuracy in reporting of trip length. Some respondents reported a multi-trip journey as one trip rather than providing details about the separate legs of the journey, consequently there was some evidence of inaccurate recording of start and end times of trips with times given reflecting time of departure from home and time of return home.

With respect to gender, in keeping with the National Travel Survey (DTLR, 2001b), men travelled a greater distance than women and made a larger average number of trips. If men make more trips than women and travel greater distance, it would appear to follow that they would be more likely to be affected by the introduction of a charge, if they are destined to or through the city centre. It is difficult though to make such sweeping statements without also considering the nature of people's patterns of travel in greater depth. In contrast to men, if women are making less trips over a shorter distance, it may be that they are travelling into the centre while men are travelling greater distances to destinations away from the central area (and out of their own neighbourhood). Therefore, if women's travel is to the centre, depending on mode used, with charging they may incur greater travel cost if travelling by car but may benefit from more reliable journey times if travelling by bus and charging results in less congestion.

With respect to age, people between 16 and 25 made the most trips, those aged between 26 and 35 and between 46 and 65 made approximately 2 less trips per day than those in the youngest age group. People aged 36-45 made 2.84 trips, almost one trip less than respondents in age groups immediately younger or older than them. The group making the least trips was those aged 66 and over. The respondents aged 26-35 travelled the greatest average distance of about 40 km, while the over 66s travelled the shortest distance of 8.13 km. It is apparent that the least mobile group in terms of both average number of trips made and distance travelled was those aged 66+. If mean distance travelled is used as the indicator of mobility, the most mobile group were those aged 26-45. However, if average number of trips made is the mobility indicator, then those in the youngest age group were the most mobile. This evidence indicates that if young people aged 16-25 make the most trips and those aged 26-35 travel the greatest distance, these groups may incur the largest increase in travel cost if a charge is introduced and if they are destined by car to or through the central area.

In terms of ethnicity, there was little difference in the mean number of trips made by both White and Asian participants at just over three trips per day. However, the average distance travelled by White respondents was double that of the Asian participants. The large standard deviation associated with the mean distance for the White group indicates that this figure may be unusually high because a few respondents travelled relatively large distances on the day reported. Nevertheless, the difference may also relate to social and employment activity patterns of the two groups and the prevalence of local working and involvement in neighbourhood activities that Asian participants in focus groups and interviews had alluded to, in contrast to a more diverse pattern of participation amongst White respondents. The ten African-Caribbean respondents made an average of 2.5 trips, however, the number reporting their distance travelled was very low and thus it is difficult to draw any definite conclusion about this aspect of their travel in comparison to other ethnic groups: the four African-Caribbean participants providing trip distances travelled an average of 21.13 km, a similar distance to that travelled by the White respondents.

These findings indicate that since White and Asian people make the most trips, they could be more likely to be affected by the introduction of a charge: if travelling by car, they may experience greater travel costs but benefit from less delays and, if travelling by bus, they may benefit from less delays. However, with African-Caribbean people travelling a similar distance to White residents, they may also experience the same cost and benefit increases. Perhaps these findings on mobility more than those for any of the other groupings reported above suggest that it is difficult to isolate potential effects of road user charging without reference to issues such as destination, trip purpose and mode choice.

Table A2.2   Average Number of Trips and Distance Travelled Per Person on Research Day by Groups Studied – Bristol

Descriptive variable	Number of trips per person			Distance travelled per person (km)		
	No.	Mean	S.D.	No.	Mean	S.D.
Area of residence						
Inner	48	3.00	2.01	39	16.96	29.05
Outer	18	3.67	2.63	15	24.85	27.01
Total	66	3.18	2.20	54	19.16	28.47
Gender						
Male	30	3.50	2.24	25	22.54	32.14
Female	36	2.92	2.16	29	16.23	25.10
Age Group						
16-25	5	5.40	3.13	4	21.90	12.01
26-35	6	3.67	1.21	4	39.63	34.82
36-45	19	2.84	2.19	15	26.14	40.26
46-55	14	3.50	2.47	12	11.27	15.94
56-65	8	3.50	1.77	6	25.88	39.21
66+	14	2.14	1.61	13	8.13	7.54
Ethnic Group						
White	36	3.33	2.70	32	23.43	34.53
African-Caribbean	10	2.50	1.27	4	21.13	35.00
Asian	19	3.32	1.42	17	11.05	5.89
Other	1	2.00	-	1	12.40	-

Nevertheless, taking all these findings on mobility, it is possible that road user charging may have greater direct impacts on those people with higher levels of mobility:

- outer city residents
- males
- 16-25 and 26-35 year olds
- Asian residents

The challenges of generalizing by group about potential impacts of road user charging indicate the importance of taking account of the social features of each group's travel to help postulate possible effects. For this reason the report now examines each group's trip purposes and then goes on to describe the findings on modes of travel.

2.1.3 Gender and Trip Purpose

Table A2.3 shows most trips made by respondents on the weekdays when they completed diaries were to/for work or to education, with 32% of women's trips being for this purpose and 26% of men's. In terms of other trip purposes, there was relatively little variation between males and females (ranging from a difference of between 2% and 4% of trips made by that gender) except for trips home where 25% of trips by men were for this purpose in contrast to 20% of those made by women.

Examining the two most common purposes for each gender further to investigate the possible implications of the findings with regard to road user charging, it has been shown that trips to/for work and to education were the most common purposes for both genders. For males, trips home were the next most common and, for females, shopping or personal business trips were next. This evidence suggests that males and females are equally likely to be affected on work/education trips by the introduction of a charge (if travelling by car – increased travel cost/less delay, if travelling by bus – possibly lower journey times/greater reliability). However, since journeys home would tend to be outbound from the centre, trips made by men for their second most frequent purpose would generally not be subject to a charge. For women, with shopping and personal business trips destined to both local and city centre locations (as indicated on the travel diaries), some of the trips for their second most common purpose would result in increased travel cost if a charge was introduced (and if they were travelling by car). It would appear then that the most common types of trips made by women have greater potential to be subjected to a cordon charge.

Table A2.3   Trip purpose by Gender – Bristol

Purpose	Male	%	Female	%	Total
To go to work/ education/for work	27	26	32	32	59
To go home	26	25	20	20	46
To take a passenger/someone somewhere	5	5	7	7	12
Social/entertainment	17	16	12	12	29
Shopping or personal business	25	24	21	21	46
Other	6	6	8	8	14
Total	106	102	100	100	206

* % of trip purposes in gender

2.1.4 Age Group and Trip Purpose

In terms of age group, Table A2.4 illustrates that for people aged 56 and over, the most frequent trip purpose was shopping or personal business. In all the other age groups, trips to/for work and to education made up the largest proportion of daily trips (ranging from 25% of trips made by 16-25 year olds to 50% made by 26-35 year olds). Travelling to take someone somewhere (eg. a child to school/spouse to work) was most commonly seen amongst 36-45 year olds and not described at all by those aged 16-35 or those aged 66 and over. Thirty per cent of the trips made by people over 66 were for social and entertainment purposes. The other age group with a relatively high proportion (21%) of social and entertainment trips was the 16-25 year olds. In contrast, none of the trips made by people aged 26-35 were for this purpose.

With particular reference to people studied at the two ends of the lifecycle spectrum, the most common trip purpose for those aged 16-55 was to/for work or to education and, for those aged 55+, was for shopping or personal business. Trips of both these types are equally likely to be local as they are to be destined to the city centre (and the raw travel diary data indicated this combination of geographic destinations). As such, people in these two age groups may be affected by a charge when travelling for their most frequent trip purpose to destinations to/through the centre by car. Conversely, with decreased congestion that may be associated with road user charging and hypothecated revenue applied to improvements in bus services, both age groups may also benefit from charging when not travelling by car for their main trip purpose.

Table A2.4   Trip Purpose by Age Group – Bristol

Purpose	16-25	%*	26-35	%	36-45	%	46-55	%	56-65	%	66+	%	Total
To go to work/education/ for work	6	25	11	50	20	38	17	35	4	15	1	3	59
To go home	5	21	6	27	11	21	12	25	7	26	5	17	46
To take a passenger/some one somewhere	0	0	0	0	8	15	2	4	2	7	0	0	12
Social/ entertainment	5	21	0	0	5	9	6	12	4	15	9	30	29
Shopping or personal business	4	17	3	14	6	11	10	20	8	30	15	50	46
Other	4	17	2	9	3	6	2	4	2	7	0	0	13
Total	24	101	22	100	53	100	49	100	27	100	30	100	205

* % of trip purposes in age group

2.1.5 Ethnic Group and Trip Purpose

It is difficult to make comparisons between African-Caribbean trip-making patterns and those of the White and Asian groups because of the relatively small number of diaries completed by African-Caribbean respondents, thus any findings related to this ethnic group must be treated with caution. Nevertheless, Table A2.5 shows that the most common trip purpose amongst White and African-Caribbean respondents was to go to or for work or to education whereas the most frequent trip purpose amongst the Asian respondents was shopping or personal business. In the White and Asian respondents' trips, similar levels of trip-making for the purposes of going home (23% and 24% respectively) and social/entertainment (15% and 16% respectively) were found.

Work/education trips (the most common trip purpose for both White and African-Caribbean ethnic groups) and shopping or personal business trips (the most common for Asian respondents) were destined to various geographic locations with travel diaries describing work/education and shopping/personal business destinations both inside and outside the central area. It is clear then that all ethnic groups may incur the charge if travelling by car for their most common trip purpose, likewise if using other modes such as walking or bus for their most frequent trip purpose, there is potential for all groups to benefit from hypothecated revenue investment and lowering of congestion levels.

Table A2.5   Trip Purpose by Ethnic Group – Bristol

Purpose	White	%*	African-Caribbean	%	Asian	%	Other	%	Total
To go to work/ education/for work	38	32	11	46	10	16	0	0	59
To go home	27	23	3	13	15	24	1	50	46
To take a passenger/ someone somewhere	9	8	1	4	2	3	0	0	12
Social/ entertainment	18	15	0	0	10	16	1	50	29
Shopping or personal business	17	14	5	21	24	38	0	0	46
Other	9	8	4	17	2	3	0	0	15
Total	118	100	24	101	63	100	2	100	207

* % of purposes in ethnic group

2.1.6 Gender and Main Mode

Table A2.6 shows that the dominant mode of travel for males was driving a car while for women walking was the most frequently used mode. In contrast to women whose second most common mode was the bus, men's second most frequently used mode was travelling as a car passenger. The third most frequent mode for women was car driving and, for men, using the bus or walking.

In other words, the two most common modes used for trips by males were car-based (as a driver: 40%, as a passenger: 17%) while, for women, walking was most common (40%) followed by bus (24%). This evidence suggests that if trips are to/through the centre, men are most likely to incur the charge, although they could then also be expected to gain most from direct benefits such as lower traffic levels that may result. With hypothecation, women's trips by bus are probably most likely to be improved, although with such high levels of walking reported by women, there is a clear need for some of the revenue to be directed towards improving the street environment, pedestrian facilities and general enhancements that increase the amenity of the walking experience.

It must also be noted that car driving is the third most common mode used by women, so charging would directly impact some women's travel, for example, since women have been seen to make shorter trips, a charge for a short journey by car to the city centre would be a much greater proportion of their travel cost for that trip than for a man making a trip from a greater distance to the same geographic destination. Taking this further, with women making more escort trips, if such a journey is to the centre, the direct effect of an increased trip cost would also affect the passenger (and possibly the woman's ability to provide the lift for the passenger).

Table A2.6   Main Mode Used by Gender – Bristol

Mode	Male	%*	Female	%	Total
Car/van driver	42	40	17	17	59
Motorcycle/moped	8	8	0	0	8
Car/van passenger	18	17	13	13	31
Goods vehicle	0	0	0	0	0
Pedal cycle	4	4	2	2	6
Walk	14	14	40	40	54
Bus	15	14	24	24	39
Taxi	0	0	0	0	0
Train	2	2	3	3	5
Other	1	1	0	0	1
Total	104	100	99	100	203

* % of main mode used in gender

2.1.7 Age Group and Main Mode

Table A2.7 shows that, amongst those aged 16-25, the dominant modes were walking and travelling as a car/van passenger. For those aged 66 and over, travelling by car was dominant with 46% of trips made as a passenger and 32% as a driver. In all other age groups, car driving was most frequent except amongst 46-55 where this was usurped by walking (6% more trips in this age group). Bus use was the second most common mode for 56-65s, 46-55s (tied with car driving) and 26-35s. For 36-45 year olds, walking was the second most dominant mode.

Looking more closely at the travel patterns of the elderly (66+), the evidence indicates how car-dependent this group is. Although people in this age group make the least number of trips, it would appear that most of these are by car and therefore liable to charging, depending on trip destination. In light of elderly people's lower incomes, a charge would be another payment to be made from what are often constrained household budgets. In addition, some elderly people find public transport difficult to use through disability or, as revealed in the focus group research, the debilitating effects of illness and treatment, which indicates that for some car-based travel may be the only feasible mode available. For this group, financial and medical constraints may mean that road user charging substantially lowers their accessibility to key services.

At the other end of the age range, people aged 16-25 tended to walk (39%) or travel as car/van passengers (35%). This age group reported making only 15% of trips by bus. It can be inferred that bus services do not satisfy this group's travel needs with travel as a car passenger seeming to substitute for the deficit. If trips to the city centre are being made as a passenger in a car, will the lifts still be available if travel is more expensive when a charge is introduced? Or will hypothecated revenue fill the gaps in bus services and make them more attractive generally to this group (some of whom described buses as rather unsavoury and insecure at focus groups), so that such a widespread reliance on lifts will cease to be necessary?

Table A2.7   Main Mode Used by Age Group – Bristol

Mode	16-25	%*	26-35	%	36-45	%	46-55	%	56-65	%	66+	%	Total
Car/van driver	0	0	9	43	19	37	13	27	9	33	9	32	59
Motorcycle/ moped	3	12	0	0	4	8	0	0	0	0	0	0	7
Car/van passenger	9	35	1	5	1	2	5	10	3	11	13	46	32
Goods vehicle	0	0	0	0	0	0	0	0	0	0	0	0	0
Pedal cycle	0	0	0	0	5	10	1	2	0	0	0	0	6
Walk	10	39	3	14	13	25	16	33	7	26	5	18	54
Bus	4	15	6	29	5	10	13	27	8	30	1	4	37
Taxi	0	0	0	0	0	0	0	0	0	0	0	0	0
Train	0	0	2	10	3	6	0	0	0	0	0	0	5
Other	0	0	0	0	1	2	0	0	0	0	0	0	1
Total	26	101	21	101	51	100	48	99	27	100	28	100	201

* % of main mode used in age group

2.1.8 Ethnic Group and Main Mode

Once again, the provisos attached to interpretation of African-Caribbean data apply. However, Table A2.9 indicates that for both White and Asian respondents, car driving was the dominant mode. In contrast, walking was the mode most frequently used for trips made by African-Caribbean respondents. For trips made by White respondents, walking was the second most common mode and travelling as a car passenger third. For trips made by African-Caribbean respondents, bus was the second most frequent mode, followed by car passenger and train. For Asian trips, car passenger and walking trips were in joint second place.

As has previously been stated, White and Asian respondents made trips most frequently as car drivers. This reliance on a car/van for trip-making is particularly marked amongst the Asian group where travelling as a car/van passenger was second most frequent (tied with walking at 21%). So, for trips made by White respondents, 51% were in a car (both driver and passenger trips) and, for trips by Asian respondents, 59% were by car. If

Asian residents are, as the evidence implies, most likely to be affected by road user charging because of high levels of car/van dependence, it follows that they could be most vulnerable to exclusionary effects of the charge (eg. the Asian elderly who described an unfamiliarity with bus use at focus groups may end up forfeiting journeys if they cannot afford increased travel cost associated with a cordon charge and are unable to use the bus services). This indicates the need for hypothecated revenue spending to be directed towards educational and marketing campaigns in appropriate languages to make public transport alternatives more accessible to this population.

Examining the car-based trips made by Asian respondents in terms of gender, an interesting pattern is revealed. Trips as a car driver and as a car passenger by each gender were compared (Table A2.8). Four-fifths of the trips made by Asian men were as car drivers in contrast to Asian women who made approximately a quarter of their trips by this mode, relying on travel as a car passenger for three in every four trips instead. It has already been suggested that the Asian population in Bristol could potentially be more vulnerable to direct cost impacts of road user charging. However, it is also apparent from this comparative mode choice pattern that there may also be gender differences in vulnerability to impacts within this ethnic group as the dependence of Asian women on lifts may increase their exposure to deleterious effects of a charge since increased travel costs could mean that lifts become less readily available because of financial constraints on family budgets. Consequent suppression of trips or enforced mode change may follow.

Table A2.8   Number of Car-based Trips by Asian Respondents by Gender

Main Mode	Male Trips (% of all car-based trips)	Female Trips (% of all car-based trips)
Car Driver	20 (56%)	3 (8%)
Car Passenger	5 (14%)	8 (22%)

The African-Caribbean respondents reported that 44% of their trips were on foot, 26% by bus and 9% as a car/van passenger. It is interesting to highlight that conversation with African-Caribbean residents of Easton indicated just how much walking activity takes place on a daily basis. There was no apparent stigma attached to walking in this area because everyone is doing it. Walking was seen as acceptable – it is a social equalizer and, in light of views on negative views expressed by the same residents about buses, there was an apparent desirability in walking. This was countered by a feeling amongst those who walked mainly that car drivers should not be penalized with road user charging as this would be unfair since they suffer/pay enough already.

The evidence suggests that the ethnic group that may be most affected by the introduction of a charge is Asian, followed closely by White. With African-Caribbean people reporting the highest percentage of trips by bus amongst the three ethnic groups, there is a possibility that they could make the most gains from hypothecation if this enhances bus services.

Table A2.9   Main Mode Used by Ethnic Group – Bristol

Mode	White	%*	African-Caribbean	%	Asian	%	Other	%	Total
Car/van driver	35	36	1	4	23	38	0	0	59
Motorcycle/moped	8	8	0	0	0	0	0	0	8
Car/van passenger	15	15	2	9	13	21	1	50	31
Goods vehicle	0	0	0	0	0	0	0	0	0
Pedal cycle	5	5	1	4	0	0	0	0	6
Walk	31	32	10	44	13	21	0	0	54
Bus	1	1	6	26	12	20	1	50	20
Taxi	0	0	0	0	0	0	0	0	0
Train	3	3	2	9	0	0	0	0	5
Other	0	0	1	4	0	0	0	0	1
Total	98	100	23	100	61	100	2	100	184

* % of main mode used in ethnic group

2.2 Research findings – Nottingham

2.2.1 Travel Diary Distribution and Return

Table A2.10 illustrates how diaries were distributed in Nottingham and the rate of return of questionnaires.

Table A2.10 Travel Diary Distribution and Return – Nottingham

Travel Diary Distribution and Receipts	Number
Number of travel diaries distributed @ focus groups	67
Number of travel diaries distributed @ interviews	10
Number distributed to disaffected youth club	4
Number distributed to young women's group	6
Number of travel diaries distributed by 'snowball'	59
Total number of diaries distributed	**146**
Number of travel diaries returned	**71**
Rate of return	**46%**
Number of returned diaries providing 'snowball' contacts	34
% of respondents providing 1 or more 'snowball' contacts	48%

There was a response rate of almost 50% to the travel diary survey in the city. This was an acceptable rate for diaries distributed through face-to-face contact. The difficulty experienced in Nottingham was in making contact with African-Caribbean groups in the city (described in Interim Report: Focus Groups) and this contributed to a lower number of diaries being distributed than projected at inception of the project and, consequently, led to an overall lower number of returns than initially intended.

In terms of socio-demographic background, 65% of respondents were female, 4% were 25 and under and 32% 66 and over, 25% were from an ethnic minority.

2.2.2 Average number of trips and distance travelled

Respondents were classified by area of residence into inner and outer wards. Table A2.11 shows that residents of outer city wards were less mobile than those in inner city wards with the former travelling an average distance of 14.88 km in an average of 3.20 trips and the latter 20.36 km in 4.38 trips. This pattern of reverse mobility was unexpected and may be explained by the fact that many of the respondents from outer Nottingham were from Clifton which is somewhat isolated from the main built-up area and relatively self-contained. Analysis of all trips to/for work by car drivers indicates that 85% of these trips were made by inner city ward residents. This probably relates to the difference in the ages of respondents from the two geographic areas: 46% of trips reported by respondents in the outer wards were from the oldest age category (who would tend not to be making work trips). Indeed, 25% of trips made by inner ward residents were made by the most mobile group (36-45 year olds), indicating that comparisons of potential relative levels of exposure to workplace parking levy based on area of residence would not be useful for this sample.

The average length of trip reported in Nottingham was 32 minutes with a minimum length of 1 minute and a maximum of 505. The higher figure may be inflated as examination of individual diaries suggested that some respondents were reporting times of departure from and return home rather than individual trip timings.

In terms of gender, men travelled an average of three times the distance of women. Therefore, in terms of trips made, men were more mobile than women. However, the average numbers of trips made by each of these two groups were not as diverse with males making 0.57 more trips than females.

People aged between 46 and 55 made the highest average number of trips (5.11) – double the number of trips made by the over 66s. Respondents in all age groups except the eldest, made 4 or 5 trips a day. In terms of distance travelled, the 36-45 year olds showed the largest average distance, however this figure was associated with a high standard deviation which may indicate that was unusually high because of an individual reported an exceptionally long trip in contrast to others in the group. People aged over 66 travelled the least distance.

Looking at both measures of mobility, with 2.55 trips and 7.36 km travelled, the over 66s were the least mobile age group. The most mobile age group in terms of number of trips made was the 16-25 year olds and, in terms of distance travelled, the 36-45 year olds.

It is impossible to make any deduction on trip-making patterns of the African-Caribbean ethnic group as only one respondent provided detailed information on the number of trips made and distance travelled. In terms of White and Asian, the number of trips made by White participants were slightly higher at 3.89 than those made by Asian at 3.19. Nevertheless, the similarity in number of trips made is not reflected in the average distance travelled with White respondents reporting an average of double the distance of Asians. If levels of mobility are used as an indicator of probability that workplace parking levy may affect one group more than another, it would appear that White residents may be exposed to a greater effect. However, if this group travels more, it is also likely to benefit most from any potential gains associated with the levy such as decreased congestion and investment of hypothecated revenue in transport improvements.

Taking all these findings on mobility, it is postulated that workplace parking levy may have greater direct impacts on those people with higher levels of mobility:

- inner city residents (though it is problematic determining how accurate a reflection of comparative mobility levels by area of residence descriptor has been revealed by the research)
- males
- 16-25 and 36-45 year olds
- White residents

However, how much workplace parking levy affects any of the people in these groups can really only be surmized when travel mode and trip purpose are also examined.

Table A2.11 Average Number of Trips and Distance Travelled Per Person on Research Day by Groups Studied – Nottingham

Descriptive variable	No of trips per person			Distance travelled per person (km)		
	No.	Mean	S.D.	No.	Mean	S.D.
Area of residence						
Inner	40	4.38	2.48	40	20.36	43.99
Outer	30	3.20	2.23	30	14.88	21.95
Total	71	3.83	2.44	71	17.76	35.91
Gender						
Male	25	4.20	2.80	25	30.34	57.72
Female	46	3.63	2.22	46	10.92	9.65
Age Group						
16-25	3	5.00	2.65	3	12.87	7.83
26-35	11	4.27	2.00	11	14.16	12.52
36-45	13	4.38	2.57	13	43.34	78.07
46-55	9	5.11	3.02	9	18.68	11.53
56-65	11	4.36	2.34	11	15.42	11.10
66+	22	2.55	1.95	22	7.36	8.30
Ethnic Group						
White	53	3.89	2.48	53	20.13	41.06
African-Caribbean	1	7.00	-	1	20.10	-
Asian	16	3.19	2.04	16	10.12	9.47
Other	1	3.83	-	1	12.00	-

2.2.3 Gender and Trip Purpose

Table A2.12 shows that most trips in this city were for the purpose of going home. Amongst trips made by males, this purpose was equalled in frequency by trips to/for work and to education. In contrast, shopping or personal business trips were almost as common as trips home for female respondents. Men made a larger proportion of social and entertainment trips while women made a slightly higher percentage of trips to take someone somewhere. In terms of workplace parking levy, only trips to work will be directly affected, therefore 26% of those made by men and 22% of those made by women would potentially be subjected to some effect by the charge, if these trips were as a car driver.

Table A2.12 Trip Purpose by Gender – Nottingham

Purpose	Male	%*	Female	%	Total
To go to work/education/for work	27	26	36	22	63
To go home	27	26	49	30	76
To take a passenger/someone somewhere	5	5	11	7	16
Social/entertainment	17	16	12	7	29
Shopping or personal business	22	21	48	29	70
Other	7	7	9	5	16
Total	105	101	165	100	270

* of trip purpose in gender

2.2.4 Age Group and Trip Purpose

It can be assumed that people of working age would be most affected by a levy on parking at the workplace. However, the evidence suggests that levels of trip-making to work vary amongst this large group of people. Trips to/for work or to education were the most common purpose for those aged 26-35 and 36-45. For those aged 46-55, travelling home was most frequent and, amongst 56-65 year olds, shopping/personal business trips were most common. People in the youngest age category made most of their trips to home and made an equal number of trips to/for work or education as for shopping or personal business. For people at the other end of the age range (66+), and largely outside of the work environment, almost half their reported trips were for shopping or personal business with trips home being the next most frequent.

The findings suggest that workplace parking levy would potentially have greatest impact on people aged 26-45 (with the provisos that they are travelling by car and parking at a workplace where charging is applicable and passed on to employees).

Table A2.13 Trip Purpose by Age Group – Nottingham

Purpose	16-25	%*	26-35	%	36-45	%	46-55	%	56-65	%	66+	%	Total
To go to work/education/for work	3	20	16	34	22	39	11	24	5	11	5	9	62
To go home	7	47	13	28	17	30	13	28	11	23	15	28	76
To take a passenger/ someone somewhere	1	7	5	11	3	5	0	0	7	15	0	0	16
Social/ entertainment	1	7	7	15	2	3	2	4	11	23	6	11	29
Shopping or personal business	3	20	5	11	10	18	12	26	12	26	26	48	68
Other	0	0	1	2	3	5	8	17	1	2	2	4	15
Total	15	101	47	101	57	100	46	99	47	100	54	100	266

* of trip purpose in age group

2.2.5 Ethnic Group and Trip Purpose

With such a low number of trips reported by African-Caribbean respondents, it is clear that any interpretation of the statistical evidence must be treated with caution. This does not preclude some explanation of the research findings and, in particular, those for the White and Asian groups. In Table A2.14 it can be seen that, for trips made by White respondents, the greatest proportion were to go home and, for trips made by Asian respondents, the most common purpose was shopping or personal business. It can be noted too that for the African-Caribbean trips, the most frequent purpose was to go home. With respect to the second and third most common trip purposes reported, for trips made by Whites, shopping or personal business and work and education trips followed respectively and, for trips made by Asians, work and education and then social and entertainment trips followed in frequency.

Making trips to/for work or to education was the second most common purpose for Asian respondents (25%) and third most common for White respondents (23%). In other words, the workplace parking levy has the potential to affect a quarter of the trips made by these two ethnic groups. For trips by people from the African-Caribbean community, workplace parking levy may affect 14% of trips.

TableA 2.14 Trip Purpose by Ethnic Group – Nottingham

Purpose	White	%*	African-Caribbean	%	Asian	%	Other	%	Total
To go to work/ education/for work	47	23	1	14	13	25	2	25	63
To go home	62	30	3	43	9	18	2	25	76
To take a passenger/ someone somewhere	13	6	2	29	1	2	0	0	16
Social/ entertainment	15	7	0	0	12	24	2	25	29
Shopping or personal business	52	26	1	14	15	29	2	25	70
Other	15	7	0	0	1	2	0	0	16
Total	204	99	7	100	51	100	8	100	270

* of trip purpose in ethnic group

2.2.6 Gender and Main Mode

Table A2.15 shows that a large proportion of the trips by males were as car/van drivers (61%) making this their most commonly used mode. In comparison, 31% of trips made by females were as car/van drivers (second most frequently reported mode for trips by this gender) and, instead, walking was the most frequently reported mode (33%). For men, walking was the second most common mode and the bus came third. Amongst trips made by women, bus was the third most frequently used mode.

With regard to workplace parking levy, the only mode directly affected would be car driving. Therefore, in terms of gender, 61% of men's trips and 31% of women's could be affected. However, indirect impacts such as improved traffic flow due to less congestion after policy introduction could have impacts on all other road-based modes. So, for the 13% of trips by males and the 24% of females made by bus, there would be potential for journey time improvements and greater service reliability. Lower congestion levels could also be beneficial for people walking on pavements by busy roads (18% of trips by men and 33% of those made by women were on foot, though it is impossible to determine from the data how many were on- or off-road).

Table A2.15 Main Mode Used by Gender – Nottingham

Mode	Male	%*	Female	%	Total
Car/van driver	64	61	51	31	115
Motorcycle/moped	2	2	1	1	3
Car/van passenger	0	0	5	3	5
Goods vehicle	0	0	0	0	0
Pedal cycle	3	3	8	5	11
Walk	19	18	55	33	74
Bus	14	13	40	24	54
Taxi	3	3	2	1	5
Train	0	0	0	0	0
Other	0	0	4	2	4
Total	105	100	166	100	271

*% of main mode used in gender

2.2.7 Age Group and Main Mode

Over 40% of all trips reported were made by car and, therefore, potentially liable for levy payment. However, this figure must be higher than the actual number of trips that would be affected by the charge since, for instance, it includes trips made by people aged 66 and over who tended not to make trips to work and young people who may have been making trips to education rather than work and would not have to pay a charge. The 16-25 year olds' trips were most frequently made as car/van driver with walking being the next most common mode and bus third. For the over 66s, bus was most common followed by walking and then driving a car/van. Amongst trips made by 26-35 year olds, walking was the most frequently reported mode followed very closely by car/van driving and then by cycling. For the other three age groups, the most common mode was driving a car/van and this dominance was particularly marked for 36-45 year olds where the mode made up 63% of trips and for 56-65 year olds where it was used for 55% of trips. For 46-55 year olds, there was much less dominance of one mode with walking being reported almost as frequently as car/van driving.

It follows then that the data suggest that the age groups who made the largest proportion of trips by car were 36-45 and 56-65 year olds (63% and 55% respectively) and, thus, in terms of age categories, have the greatest potential to be affected by a levy.

In addition, it is worthwhile reiterating that people in all age groups would gain from any attendant congestion decrease associated with the levy and, in terms of bus users, this may be of particular relevance for the oldest age group where 33% of trips (the highest proportion in all age groups and most frequently used mode within this group) were made by this mode.

Table A2.16 Main Mode Used by Age Group – Nottingham

Mode	16-25	%*	26-35	%	36-45	%	46-55	%	56-65	%	66+	%	Total
Car/van driver	8	53	13	28	36	63	17	37	26	55	15	27	115
Motorcycle/ moped	0	0	0	0	2	4	0	0	0	0	0	0	2
Car/van passenger	0	0	0	0	2	4	2	4	0	0	1	2	5
Goods vehicle	0	0	0	0	0	0	0	0	0	0	0	0	0
Pedal cycle	0	0	11	23	0	0	0	0	0	0	0	0	11
Walk	5	33	14	30	6	11	16	35	15	32	17	31	73
Bus	2	13	8	17	11	19	8	17	5	11	18	33	52
Taxi	0	0	1	2	0	0	3	7	1	2	0	0	5
Train	0	0	0	0	0	0	0	0	0	0	0	0	0
Other	0	0	0	0	0	0	0	0	0	0	4	7	4
Total	15	100	47	100	57	101	46	100	47	100	55	100	267

* % of main mode used in age group

2.2.8 Ethnic Group and Main Mode

The previously-described need for prudence with regard to the analysis of African-Caribbean trip-making patterns must be underlined. The most frequently used mode for White, Asian and African-Caribbean respondents was car driving. With few trips reported by the African-Caribbean group, no further comment can be made. Nevertheless, since 55% of trips made by Asian respondents and 40% of those made by White respondents were as car drivers, it is possible to deduce that the direct impacts of a levy may potentially be greater on the Asian than the White community. Interestingly, for both these ethnic groups, walking was the next most frequently used mode, indicating that it may be desirable to ring-fence a substantial proportion of hypothecated revenue for improvements that enhance the pedestrian environment.

Table A2.17 Main Mode Used by Ethnic Group – Nottingham

Mode	White	%*	African-Caribbean	%	Asian	%	Other	%	Total
Car/van driver	82	40	5	71	28	55	0	0	115
Motorcycle/ moped	3	1	0	0	0	0	0	0	3
Car/van passenger	5	2	0	0	0	0	0	0	5
Goods vehicle	0	0	0	0	0	0	0	0	0
Pedal cycle	11	5	0	0	0	0	0	0	11
Walk	52	26	2	29	16	31	4	50	74
Bus	44	22	0	0	6	12	4	50	54
Taxi	4	2	0	0	1	2	0	0	5
Train	0	0	0	0	0	0	0	0	0
Other	4	2	0	0	0	0	0	0	4
Total	205	100	7	100	51	100	8	100	271

% of main mode used in ethnic group

1 Researchers travelled by taxi between Stapleton Road in Easton and St Paul's in Ashley – a journey of about 3 minutes – because it was raining heavily. This cost £3.20 for a distance that could easily have been walked if it were not that the walking route takes you alongside a busy dual carriageway to a roundabout over the M32. The route is austere and not conducive to walk or cycle journeys. Yet it was reported by research participants that they had familial ties in these areas but that it is not possible to travel by bus between these two areas and taxi fares were high.

Appendix 3

Policy Guidance Concerning Application of Title VI of the Civil Rights Act of 1964 to Metropolitan and Statewide Planning

(Extract from http://www.epa.gov/fedrgstr/EPA-IMPACT/2000/ May/Day-19/i12590.htm accessed 08 Nov 2003)

[Federal Register: May 19, 2000 (Volume 65, Number 98)]
[Rules and Regulations]
[Page 31803-31805]
From the Federal Register Online via GPO Access [wais.access.gpo.gov]
[DOCID:fr19my00-5]

[[Page 31803]]

DEPARTMENT OF TRANSPORTATION

Federal Highway Administration
23 CFR Parts 450 and 771
49 CFR Parts 619 and 622
Federal Transit Administration

Policy Guidance Concerning Application of Title VI of the Civil Rights Act of 1964 to Metropolitan and Statewide Planning

AGENCIES: Federal Highway Administration (FHWA), and Federal Transit Administration (FTA), DOT.

ACTION: Notice of policy.

SUMMARY: This document publishes guidance regarding the implementation of Title VI of the Civil Rights Act of 1964 (42 U.S.C. 2000d-2000d-4) concerning nondiscrimination in federally assisted programs, in metropolitan and statewide planning. This guidance was previously issued on October 7, 1999, as a memorandum to FTA Regional Administrators and FHWA Division Administrators, and is printed in its entirety.

FOR FURTHER INFORMATION CONTACT: For application to metropolitan planning, Mr. Sheldon M. Edner, FHWA, (202) 366-4066 or Mr. Charles Goodman, FTA, (202) 366-1944. For application to statewide planning, Mr. Dee Spann, FHWA, (202) 366-4086 or Mr. Paul Verchinski, FTA, (202) 366-1626. All are located at the U.S. Department of Transportation, 400 Seventh Street, SW., Washington, DC 20590-0001. Office hours are from 7:45 a.m. to 4:30 p.m., e.t., Monday through Friday, except Federal holidays.

SUPPLEMENTARY INFORMATION:

Electronic Access

An electronic copy of this document may be downloaded by using a computer, modem and suitable communications software from the Government Printing Office's Electronic Bulletin Board Service at (202) 512-1661. Internet users may reach the Office of the Federal Register's home page at http://www.nara.gov/fedreg and the Government Printing Office's web page at http://www.access.gpo.gov/nara.
(Authority: 23 U.S.C. 315; 49 CFR 1.48 and 1.51)

Issued on: May 9, 2000.
Nuria I. Fernandez,
Acting Administrator.

Kenneth R. Wykle,
Federal Highway Administrator.

The guidance memorandum reads as follows:

Date: October 7, 1999.
Subject: ACTION: Implementing Title VI Requirements in Metropolitan
          and Statewide Planning
From: Gordon J. Linton, Administrator, FTA
          Kenneth R. Wykle, Administrator, FHWA
To: FTA Regional Administrators
          FHWA Division Administrators

Background

The purpose of this memorandum is to issue clarification to you in implementing Title VI of the 1964 Civil Rights Act (42 U.S.C. 2000d-1) and related regulations, The President's Executive Order on Environmental Justice, the U.S. DOT Order, and the FHWA Order.

Title VI states that "No person in the United States shall, on the ground of race, color, or national origin, be excluded from participation in, be denied the benefits of, or be subjected to discrimination under any program or activity receiving Federal financial assistance." Title VI bars intentional discrimination as well as disparate impact discrimination (i.e., a neutral policy or practice that has a disparate impact on protected groups).

The Environmental Justice (EJ) Orders further amplify Title VI by providing that "each Federal agency shall make achieving environmental justice part of its mission by identifying and addressing, as appropriate, disproportionately high and adverse human health or environmental effects of its programs, policies, and activities on minority populations and low-income populations."

Increasingly, concerns for compliance with provisions of Title VI and the EJ Orders have been raised by citizens and advocacy groups with regard to broad patterns of transportation investment and impact considered in metropolitan and statewide planning. While Title VI and EJ concerns have most often been raised during project development, it is important to recognize that the law also applies equally to the processes and products of planning. The appropriate time for FTA and FHWA to ensure compliance with Title VI in the planning process is during the planning certification reviews conducted for Transportation Management Areas (TMAs) and through the statewide planning finding rendered at approval of the Statewide Transportation Improvement Program (STIP).

This memorandum serves as clarification pending issuance of revised planning and environmental regulations.

Requested Action

We request that during certification reviews you raise questions that serve to substantiate metropolitan planning organization (MPO) self-certification of Title VI compliance. Suggested questions are attached. Also attached are a series of actions that could be taken to support Title VI compliance and EJ goals, improve planning performance, and minimize the potential for subsequent corrective action and complaint.

Statewide planning is also subject to the same Title VI legislative requirements as the metropolitan planning process. The FHWA division offices, jointly with FTA regional offices, should review and document Title VI compliance when making the TEA-21 required finding that STIP development and the overall planning process is consistent with the planning requirements.

In part, the purpose of asking the questions attached to this memorandum is to review the basis upon which the annual self- certification of compliance with Title VI is made. The metropolitan planning certification reviews in TMAs and STIP findings offer an opportunity to FHWA and FTA staff to verify the procedures and analytical foundation upon which the self-certification is made. If it becomes evident that the self-certification was not adequately supported, a corrective action is to be included in their certification report to rectify the deficiency.

The FHWA's and FTA's Division and Regional Administrators should involve their respective civil rights staffs in the EJ and Title VI portions of the metropolitan planning certification reviews in TMAs and statewide planning findings.

Forthcoming Planning Regulations

As you know, FHWA and FTA are preparing to revise the planning (23 CFR 450 and 49 CFR 619) and environmental (23 CFR 771 and 49 CFR 622) regulations. In these rulemakings and subsequent documents, we will propose clarifications and appropriate procedural and analytical approaches for more completely complying with the provisions of Title VI and the Executive Order on Environmental Justice. Specifically, the proposals will focus on

[[Page 31804]]

public involvement strategies for minority and low-income groups and assessment of the distribution of benefits and adverse environmental impacts at both the plan and project level.

If you have questions on metropolitan applications of this memorandum, please contact Sheldon M. Edner, Team Leader, Metropolitan Planning and Policies, FHWA, (202) 366-4066; or Charlie Goodman, Division Chief, Metropolitan Planning, FTA (202) 366-1944. On statewide applications, please contact Dee Spann, Team Leader, Statewide Planning, FHWA; (202) 366-4086; or Paul Verchinski, Chief, Statewide Planning, FTA, (202) 366-1626.

Assessing Title VI Capability – Review Questions

September 1999

Discussion of these important issues will be held as part of planning certification reviews, and the discussion will be held as part of statewide planning findings that are made as part of Statewide Transportation Improvement Program (STIP) approval. These questions are offered as an aid to reviewing and verifying compliance with Title VI requirements:

1 Overall Strategies and Goals

What strategies and efforts has the planning process developed for ensuring, demonstrating, and substantiating compliance with Title VI? What measures have been used to verify that the multi-modal system access and mobility performance improvements included in the plan and Transportation Improvement Program (TIP) or STIP, and the underlying planning process, comply with Title VI?

Has the planning process developed a demographic profile of the metropolitan planning area or State that includes identification of the locations of socio-economic groups, including low-income and minority populations as covered by the Executive Order on Environmental Justice and Title VI provisions?

Does the planning process seek to identify the needs of low-income and minority populations? Does the planning process seek to utilize demographic information to examine the distributions across these groups of the benefits and burdens of the transportation investments included in the plan and TIP (or STIP)? What methods are used to identify imbalances?

2 Service Equity

Does the planning process have an analytical process in place for assessing the regional benefits and burdens of transportation system investments for different socio-economic groups? Does it have a data collection process to support the analysis effort? Does this analytical process seek to assess the benefit and impact distributions of the investments included in the plan and TIP (or STIP)?

How does the planning process respond to the analyses produced? Imbalances identified?

3 Public Involvement

Does the public involvement process have an identified strategy for engaging minority and low-income populations in transportation decisionmaking? What strategies, if any, have been implemented to reduce participation barriers for such populations? Has their effectiveness been evaluated?

Has public involvement in the planning process been routinely evaluated as required by regulation? Have efforts been undertaken to improve performance, especially with regard to low-income and minority populations? Have organizations representing low-income and minority populations been consulted as part of this evaluation? Have their concerns been considered?

What efforts have been made to engage low-income and minority populations in the certification review public outreach effort? Does the public outreach effort utilize media (such as print, television, radio, etc.) targeted to low-income or minority populations? What issues were raised, how are their concerns documented, and how do they reflect on the performance of the planning process in relation to Title VI requirements?

What mechanisms are in place to ensure that issues and concerns raised by low-income and minority populations are appropriately considered in the decisionmaking process? Is there evidence that these concerns have been appropriately considered? Has the metropolitan planning organization (MPO) or State DOT made funds available to local organizations that represent low-income and minority populations to enable their participation in planning processes?

Guidance:

Assessing Title VI Capability--FTA/FHWA Actions

Environmental Justice in State Planning and Research (SPR) and Unified Planning Work Programs (UPWPs)

At a minimum, FHWA and FTA should review with States, MPOs, and transit operators how Title VI is addressed as part of their public involvement and plan development processes. Since there is likely to be the need for some upgrading of activity in this area, a work element to assess and develop improved strategies for reaching minority and low-income groups through public involvement efforts and to begin developing or enhancing analytical capability for assessing impact distributions should be considered in upcoming SPRs and UPWPs.

Review Public Involvement Efforts During Certification Reviews for Title VI Consistency

In many areas, room for improvement exists in public involvement processes regarding engagement of minority and low-income individuals. It is appropriate to review the extent to which MPOs and States have made proactive efforts to engage these groups through their public involvement programs. Further, FHWA and FTA should review the record of complaints or concerns raised regarding Title VI in the planning process under review. During the on-site element of the metropolitan certification review, the public involvement process, now required by statute, should make a special effort to engage and involve representatives of minority and low-income groups to hear their views regarding changes to and performance of the planning process.

Options for FHWA/FTA Metropolitan Certification Review Actions

(1) FHWA and FTA should seek to determine what, if any, processes are in place to assess the distribution of impacts on different socio-economic groups for the investments identified in the transportation plan and TIP. If the planning process has no such capability in place, there needs to be further investigation as to how the MPO is able to annually self-certify its compliance with the provisions of Title VI.

(2) If no documented process exists for assessing the distributional effects of the transportation investments in the region, the planning certification report should include a corrective action directing the development of a process for accomplishing this end. This will

[[Page 31805]]

serve to put the process on notice regarding existing requirements and prepare it for future regulatory requirements. If a minimal effort is in place, FHWA and FTA should encourage the planning process participants to become familiar with the provisions of the Executive Order on Environmental Justice and identify needed improvements based on the Order.

(3) If no formal evaluation of the public involvement process has been conducted per the requirement for periodic assessment (see 23 CFR 450.316(b)), a corrective action to conduct an evaluation should be included in the certification report. The formal evaluation should, at a minimum, assess the effectiveness of efforts to engage minority and low-income populations through the local public involvement process. If the MPO or State has conducted a public involvement evaluation, FHWA and FTA should determine whether the involvement of minorities and low-income individuals has been addressed and what strengths and deficiencies were identified. Recommended improvements or corrective actions for the certification report or STIP findings can be tied to the results of the MPO's or State's public involvement evaluation.

Consolidated Bibliography

Ahmed, S., Hussain, M. and Vournas, G. (2001) 'Consultation with "hidden" and hard-to-reach groups: methods, techniques and research practice'. LARIA (Local Authorities Research & Intelligence Association) Summit, 15 November 2001 @ http://www.laria.gov.uk/content/features/68/feat1.htm

Amin, A (2002) 'Ethnicity and the multicultural city: living with diversity', Environment and Planning A 2002, Vol. 34, 959-980

Audit Commission (2001) Development Control: Best Value Report http://www.audit-commission.gov.uk/reports/BVIR.asp?CatID=ENGLISH% 5ELG%5ELOCAL-VIEW%5EAUTHORITIES%5ELG-LV-NOTCIC&ProdID =BBBBC5AC-0F26-4370-AFFD-C8072BBA7184

Audit Commission (2002) Traffic Management: Best Value Report: http://www.audit-commission.gov.uk/reports/BVIR.asp?CatID=ENGLISH% 5ELG%5ELOCAL-VIEW%5EAUTHORITIES%5ELG-LV-BRICC&ProdID =DB8DC9CB-BF67-4F44-9BA3-931AC6FDA723

Batelle (2000) Travel Patterns of People of Color. Prepared for US Dept. of Transportation, Federal Highways Administration, 2000 available @ http://www.fhwa.dot.gov/ohim/trvpatns.pdf

BBC (2003) Message Posted on BBC Bristol Message Board 14 Sep 2003 20:10 @ http://www.bbc.co.uk/cgi-perl/h2/h2.cgi?thread=%3C1061888923-23331.6% 40forum3.mh.bbc.co.uk%3E&find=%3C1061888923-23331.6%40forum3.mh. bbc.co.uk%3E&board=talkbristol.bristolsoup1&sort=Te

BBC (2001) 'Boots employs 7,500 locally' @ http://www.bbc.co.uk/nottingham/news/2001_03/28/boots.shtml

Berthoud, R. (2000) 'Family Formation in Multi-cultural Britain: Three Patterns of Diversity', Paper 2000-34, Working Papers of the Institute for Social and Economic Research. Colchester: University of Essex, December

Beuret, K., Aslam, H., Gross, S., Osman, A. and Khan, F. (2000) 'Ethnic Minorities and Visible Religious Minorities: Their Transport Requirements and the Provision of Public Transport'. Report to the DETR, London.

Bickerstaff, K. and Walker, G. (2001) 'Participatory local governance and transport planning', Environment and Planning A 2001, Vol. 33, 431-451

Blakemore, K. (2000) 'Health and Social Care Needs in Minority Communities: An Over Problematized Issue?' Health and Social Care in the Community, 8, 1, 22-30.

Bristol City Council (2000) Bristol Transport Plan 2001/2-2005/6

Burchardt T., Le Grand, J., and Piachaud, D. (1998) Social Exclusion in Britain 1991-1995 in Social Policy and Administration, Vol. 33, No. 3, pp.227-244

Campbell, C. and McLean, C. (2002) 'Social Capital and Social Exclusion in England: African-Caribbean Participation in Local Community Networks'. In Swann, C. and Morgan, A. (Eds.) Social Capital for Health: Insights from Qualitative Research. Health Development Agency, London.

Cancian, F. (1992) 'Feminist Science: Methodologies that Challenge Inequality'. *Gender and Society*, 6, 4, 623-642

Cass, N., Shove, E. and Urry, J. (2003) 'Changing infrastructures, measuring socio-spatial inclusion/exclusion'. Final Report. Department of Sociology, Lancaster University

Centre for Economic and Social Inclusion (2002) 'Social Inclusion'. http://www.cesi.org.uk/

Centre for Urban Transport Research (2000) 'Community Impact Assessment' @ http://www.ciatrans.net/index.shtml

CfIT (2002a) 'A Brief Guide to Congestion Charging' @ http://www.cfit.gov.uk/congestioncharging/factsheets/guide/index.htm#1

CfIT (2002b) 'CfIT's initial assessment report on the 10 Year Transport Plan'. CfIT, London @ www.cfit.gov.uk/research/10year/index.htm

CfIT (2002c) 'CfIT puts bus at heart of transport delivery', Press Notice 2 December 2002 @ http://www.cfit.gov.uk/pn/021202/

Church, A., Frost, M. and Sullivan, K. (2000) 'Transport and Social Exclusion in London'. *Transport Policy*, 7, 3, 195-206

Citizens Crime Commission 0f New York (CCCC) (1985) *Downtown Safety and Economic Development*, New York: Downtown Research & Development Centre

City of Vancouver (2000) *Handheld Parking Enforcement Equipment.* Administrative Report. http://www.city.vancouver.bc.ca/ctyclerk/cclerk/000406/pe3.htm

Craig, C. (2000) The analysis of workplace parking charges and their effects on business decisions with respect to reducing traffic levels. MRes in the Built Environment, Transport and Emissions Theme Group: Promoting Change of Mode in Transport, University of Leeds

Day, G. (2002) *Transport Tracker*. Institute of Directors. March

DETR (2000) *Transport 2010: The 10 Year Plan*. DETR, London

DETR (1998) *A New Deal for Transport: Better for Everyone*. The Government's White Paper on the Future of Transport. DETR, London

DfT (2003) 'Race Equality Scheme 2003-2005' @ http://www.dft.gov.uk/stellent/groups/dft_mobility/documents/page/ dft_mobility_022396.hcsp

DfT (2001) *Road Accident Involvement of Children from Ethnic Minorities: A Literature Review*. Road Safety Research Report No. 19 London, DfT

ECOPLAN (1997) 'Combined Road Pricing / Car Park Charging System for the city of Bern'

Eisenstadt, N. and Witcher, S. (1998) 'Social Exclusion and Poverty' *Outlook: The Quarterly Journal of the National Council of Voluntary Child Care Organisations*, 1, 6-7.

Enoch, M.P. (2001) 'Arriving at a transport Utopia by using an alternative policy route'. Paper prepared for the IRNES Workshop, Manchester Metropolitan University, Nov

Enoch, M.P. (2001) 'Workplace parking charges down under', *Traffic Engineering and Control*, Vol.42, No.11, November, pp.357-360.

EUBusiness (2000) Social Exclusion in EU Member States 31 Jan

Federal Register (2000) Policy Guidance Concerning Application of Title VI of the Civil Rights Act of 1964 to Metropolitan and Statewide Planning Federal Register: May 19 (Volume 65, Number 98) @ http://www.epa.gov/fedrgstr/EPA-IMPACT/2000/May/Day-19/i12590.htm

Gerrard, B., Still, B. and Jopson, A. (2001) 'The impact of road pricing and workplace parking levies on the urban economy: results from a survey of business attitudes'. *Environment and Planning A 2001*, volume 33, pp. 1985-2002.

GLC (1986) *Women on the move: Detailed Results: Black Afro-Caribbean and Asian Women*. Greater London Council, London

Goulbourne, H. and Chamberlain, M. (Eds) (2001) *Caribbean Families in Britain and the Trans-Atlantic World*. Warwick University Caribbean Studies. Macmillan.

Government Office for London and Mayor of London (2002) 'Black and Minority Ethnic Communities Cracking Crime' @ http://www.go-london.gov.uk/bmeccc/SessionReport.pdf

Graham, D., Glaister, S. and Anderson, R. (2002) *Child pedestrian casualties in England: the effect of area deprivation*. Centre for Transport Studies, Imperial College, London

Grayling, T. (2002) 'Transport and exclusion'. *The Guardian*. 21 May

Green, Grieco and Holmes (2002) 'Archiving social practice: the management of transport boycotts' @ http://www.geocities.com/the_odyssey_group/archivingsocialpractice.html

Grieco, M. (2003a) 'Transport and social exclusion: New policy grounds, new policy options,' paper presented at the 10th International Conference on Travel Behaviour Research, Lucerne, August

Grieco, M. (2003b) 'Transport boycotts and popular resistance: a Gandhian heritage'. Class notes 5, Transportation and Society with special reference to Africa, Cornell University

Grieco M. (2002) 'Limitations of transport policy: a review.' Transport Reviews 22(4)

Grieco, M. (2000) 'Intelligent urban development: the emergence of wired government and administration' in *Urban Studies*, Vol. 37, No. 10, pp. 1719-1722

Grieco, M. and Hine, J. (2002) 'Transport, Information Communication Technology and Public Service Failure: Community Monitoring and Demand Responsive Transport'. Paper presented to National Science Foundation under auspices of STELLA network, Arlington, Virginia, January

Grieco, M. and Jones, P.M. (1994) 'A change in the policy climate? Current European perspectives on road pricing.' *Urban Studies*, Vol. 31, No. 9

Grieco, M., Pickup, L. and Whipp, R. (eds) (1989) *Gender, Transport and Employment: The Impact of Travel Constraints*. Oxford Studies in Transport. Oxford

Grieco, M., Turner, J. and Hine, J. (2000) 'Transport, employment and social exclusion: changing the contours through information technology.' *Local Work*, 2000 and available at
http://www.geocities.com/transport_and_society/newvision.html

Hamilton, K., Ryley Hoyle, S. and Jenkins, L. (1999) 'The Public Transport Gender Audit: the Research Report'. Transport Studies, University of East London.

Hedges, A. (2001) *Perceptions of congestion: report on qualitative research findings.* DfT @ http://www.dft.gov.uk/stellent/groups/dft_roads/documents/ page/dft_roads_503854-01.hcsp#P63_1655

Hiley, J. (1995) *Safer Bus Routes: Final Report* Nottingham: Nottingham City Council

Hills, J., Le Grand, J. and Piachaud, D. (Eds) (2001) *Understanding Social Exclusion.* Oxford University Press, Oxford

Hine, J. and Mitchell, F. (2001), *The Role of Transport in Social Exclusion in Urban Scotland*, Central Research Unit, Scottish Executive.

Hine, J. and Mitchell, F. (2000) 'The Role of Transport in Social Exclusion in Urban Scotland'. Final Report to the Scottish Executive, Transport Research Institute, Napier University, Edinburgh

Home Office (2002) 'Race Equality in Public Services'. @
http://www.homeoffice.gov.uk/docs/raceequ_pubserv_nov02.pdf

House of Commons (2003) 'Debate on Bus Regulation' @
http://www.publications.parliament.uk/pa/cm200203/cmhansrd/cm031021/ debtext/31021-01.htm#31021-01_spnew10

Ison, S. and Rye, T. (2002) 'Lessons from travel planning and road user charging for policy-making: through imperfection to implementation'. Prepared for *IMPRINT-EUROPE Thematic Network: "Implementing Reform on Transport Pricing: Constraints and solutions: learning from best practice"*, Brussels, 23rd-24th October

Ison, S. and Wall, S. (2002) 'Attitudes to traffic-related issues in urban areas of the UK and the role of workplace parking charges'. *Journal of Transport Geography* Vol. 10, Issue 1, March, pp.21-28

Jones, P. (1998) 'Urban road pricing: public acceptability and barriers to implementation', in Button, K. and Verhoef, E. T. (eds.) *Road Pricing, Traffic Congestion and the Environment – Issues of Efficiency and Social Feasibility*, Cheltenham, Edward Elgar

Jones, P. (nd) 'Addressing Equity Concerns in Relation to Road User Charging' @ http://www.transport-pricing.net/jonel.doc

King, A. (2003) 'And they said things could only get better . . .'. *The Telegraph* 22/09/2003 @
http://www.telegraph.co.uk/news/main.jhtml?xml=%2Fnews%2F2003%2F09% 2F22%2Fnstate222.xml

Kintrea, K. and Atkinson, R. (2001) 'Neighbourhoods and social exclusion: The research and policy implications of neighbourhood effects', Department of Urban Studies Discussion Paper, University of Glasgow

Kwan, M. (2003) 'Geovisualisation of activity-travel patterns using 3D Geographical Information Systems', paper presented at the 10th International Conference on Travel Bahaviour Research, Lucerne, August 2003 @ http://www.ivt.baum.ethz.ch/allgemein/pdf/kwan.pdf

Layton-Henry, Z. (2002) 'Transnational Communities, Citizenship and African-Caribbeans in Birmingham'. ESRC Transnational Communities Programme Working Paper WPTC-02-07
http://www.transcomm.ox.ac.uk/working%20papers/WPTC-02-07%20LaytonHenry.pdf

Lefebvre, H. (1974) *La Production de l'Espace*. Published in English as *The Production of Space*. Blackwell, Oxford, 1991

Lenoir, R. (1974) *Les Exclus*. Seuil, Paris.

Litman, T. (2003) 'Accessibility: Defining, Evaluating and Improving Accessibility'. *TDM Encyclopedia*, Victoria Transport Policy Institute, British Columbia

Litman T. (1997) *Employer Provided Transit Passes: a Tax Exempt Benefit*. Victoria Transport Policy Institute, British Columbia, Canada

Litman, T. (1996) 'Using Road Pricing Revenue: Economic Efficiency and Equity Considerations', *Transportation Research Record 1558*, Transportation Research Board (http://www.nas.edu/trb/)), 1996, pp. 24-28, also available at http://www.vtpi.org/

Lucas, K., Grosvenor, T. and Simpson, R. (2001) *Transport, the Environment and Social Exclusion*. Joseph Rowntree Foundation, York

Luxton, M. (2002) *Feminist Perspectives on Social Inclusion and Children's Well Being*. Laidlaw Foundation, Canada

Lyons, G. (2003) 'Transport and Society'. Inaugural Lecture, Arup, Bristol, 1 May, University of the West of England

May, A. (1999) 'Making the links: Car use and traffic management measures in the policy package'. Paper presented to ECMT/OECD workshop on Managing Car Use for Sustainable Urban Travel, 1-2 Dec

McCluskey, A. (1997) 'Belonging and Being Excluded'. Connected. St-Blaise, 14th May @ http://www.connected.org/is/excluded.html

McCray, T. (2000) 'Van service access in Detroit for low-income women needing pre-natal care'. Proceedings of Symposium IV on African-American Mobility Issues April 30-May 2, Tampa, Florida

McCray, T., M. Lee-Gosselin, and M.-P. Kwan (2003) 'Netting action and activity space/time: are our methods keeping pace with evolving behaviour patterns?' paper presented at the 10th International Conference on Travel Behaviour Research, Lucerne, August 2003 @
http://www.ivt.baum.ethz.ch/allgemein/pdf/mccray.pdf

Metropolitan Transportation Commission (1999) 'Transportation Blueprint for the 21st Century'. Oakland California, October/November 1999 @
http://www.mtc.ca.gov/publications/transactions/ta10-1199/ta10-1199.htm

Metz, D. (2002) 'Limitations of Transport Policy'. *Transport Reviews*, 22, 2, 134-145

Metz, D. (2000) 'Mobility of Older People and Their Quality of Life'. *Transport Policy*, 7, 2, 149-152

Modood T. and Berthoud R. (1997) *Ethnic Minorities in Britain Diversity and Disadvantage*. London. Policy Studies Institute.

MORI (2001) 'The CfIT Report 2001 Public Attitudes to Transport in England: A Survey Carried out for the Commission for Integrated Transport'. July

Morris, J. (1981). 'Urban Public Transport,' in P. Troy (ed.), *Equity in the City*. Sydney: George Allen & Unwin, pp.21-49

Murray, A. and Davis, R. (2001). 'Equity in regional Service Provision.' *Journal of Regional Science*. Vol. 41, No. 4, pp.577-600

NACRO (2003) 'Getting there: reducing crime on public transport. Community safety practice briefing'. August 2003 @ http://www.nacro.org.uk/data/briefings/nacro-2003090400-csps.pdf

National Center for Transit Research (nd) 'Terminology' @ http://www.nctr.usf.edu/clearinghouse/tdmterms.htm

National Statistics Online (2002) 'Population Estimates' www.statistics.gov.uk/census2001/pop2001/bristol_city_of_ua.asp www.statistics.gov.uk/census2001/pop2001/nottingham_ua.asp

Needham, C. (2002) 'Consultation: A Cure for Local Government?' *Parliamentary Affairs*, 55, 699-714

Nottingham City Council (2001) 'Workplace Parking Levy' http://www.nottinghamcity.gov.uk/coun/department/des/wpl/front.asp

Nottingham City Council (2001) 'A new deal for public transport in Nottingham: Public Transport Plan' @ http://www.itsnottingham.info/ptplan6.htm

Nutley, S. and Thomas, C. (1995) 'Spatial Mobility and Social Change: The Mobile and the Immobile'. Sociologia Ruralis, 25, 1, 24-39

Page, B. (2003) 'A road to nowhere'. MORI @ http://www.mori.com/pubinfo/bp_road-to-nowhere.shtml

Parkes, A., Kearns, A. and Atkinson, R. (2002) 'What Makes People Dissatisfied with their Neighbourhoods?' Urban Studies, 39, 13, 2413-2438.

Pas, E. (1996) 'Recent Advances in Activity-Based Travel Demand Modeling'. Activity-Based Travel Forecasting Conference Proceedings. US Dept. of Transportation http://tmip.fhwa.dot.gov/clearinghouse/docs/abtf/pas.stm

Pearce, N. (2001) 'A Critical Analysis of the Way that Social Exclusion is Defined in Theory and Practice'. Working Paper, Lancaster University.

Phillips, M. and Phillips, T. (1998) *Windrush: The Irresistible Rise of Multi-racial Britain*. Harper Collins, London

Philo, C. (2000) 'Social Exclusion' In Johnston, R. et al. *The Dictionary of Human Geography*. Blackwell, Oxford. http://www.xreferplus.com/entry/734732.

Pratchett, L. (1999) 'New Fashions in Public Participation: Towards Greater Democracy?', *Parliamentary Affairs* 52 (4) 616-633

Prime Minister's Strategy Unit (2003) 'The future of social exclusion: drivers, patterns and policy challenges: Summary note of presentations and discussion'. Strategic Thinkers Seminar 7th May @ http://www.number-10.gov.uk/su/social%20exclusion/downloads/summary.pdf

Rajé, F. (2003) 'Impact of Road User Charging/Workplace Parking Levy on Social Inclusion/Exclusion: Gender, Ethnicity and Lifecycle Issues. Final Report'. Transport Studies Unit, University of Oxford

Rajé, F. (2002) 'Impact of Road User Charging/Workplace Parking Levy on Social Inclusion/Exclusion: Gender, Ethnicity and Lifecycle Issues. Interim Report: Focus Groups'. Transport Studies Unit, University of Oxford

Renewal.net (2002) 'Poor transport links' @ http://www.renewal.net/

Renewal.net (2002) 'Crime and transport' @ http://www.renewal.net/

Renewal.net (2002) 'Black and minority ethnic groups' @ http://www.renewal.net/

Root, A. (Ed) (2003) *Delivering Sustainable Transport: A Social Science Perspective*. Pergamon, Oxford

Root, A. (1999) 'Reconciling Environmental and Social Concerns: Transport'. A Report for the Joseph Rowntree Foundation. Transport Studies Unit, Oxford University

Root, A. (1998) 'Home Alone: Is Travel Poverty the Cinderella of Social Exclusion?' Transport Report, 21, 11. Transport 2000, London

Samers, M. (1998) 'Immigration, "Ethnic Minorities" and "Social Exclusion" in the European Union: A Critical Perspective'. *Geoforum*, 29, 123-44

Sampson (2000) *Electronic road user charging in the UK: what? why? how? and when?* DETR, London

Sánchez, Thomas W., Stolz, Rich and Ma, Jacinta S. (2003) *Moving to Equity: Addressing Inequitable Effects of Transportation Policies on Minorities*. Cambridge, MA: The Civil Rights Project at Harvard University. @ http://www.civilrightsproject.harvard.edu/research/transportation/call_trans.php

Sibley, D. (1995) *Geographies of Exclusion: Societies and Differences in the West*. Routledge, London

Sibley, D. (1981) *Outsiders in Urban Societies*. Blackwell, Oxford

Silburn, R., Lucas, D., Page, R. and Hanna, L. (1999) *Neighbourhood images in Nottingham: Social cohesion and neighbourhood change*. Joseph Rowntree Foundation @ http://www.jrf.org.uk/knowledge/findings/housing/489.asp)

Sim, J. (1998) 'Collecting and analysing qualitative data: issues raised by the focus group'. *Journal of Advanced Nursing*, 28, 2, 345-352

Sinclair, F. (2002) 'Assessment of the Effects of Congestion Charging on Low Income Households in Edinburgh - An Analysis of Scottish Household Survey Data'. PROGRESS. TRI, Napier University, January

Sinclair, F. (2001) 'Assessment of the Effects of Road User Charging and the Transport Investment Package Proposals on Social Inclusion - Recommendations for Consultation and Appraisal'. PROGRESS. TRI, Napier University, July 2001

Sinclair, S. and Sinclair, F. (2001) 'Access All Areas? An Assessment Of Social Inclusion Measures In Scottish Local Transport Strategies'. Centre for Research into Socially Inclusive Services (CRSIS) Edinburgh College of Art / Heriot Watt University, Edinburgh @ http://www.crsis.hw.ac.uk/Transport%20Review.pdf

Smyth (2000) 'The Implications of Segregation for Transport within Northern Ireland'. Ove Arup and Partners. Community Relations Council for Northern Ireland

Social Exclusion Unit (2003) 'Making the Connections: Final Report on Transport and Social Exclusion'. Office of the Deputy Prime Minister, London

Social Exclusion Unit (2001a) 'Preventing Social Exclusion'. Cabinet Office. London
Social Exclusion Unit (2001b) 'National Strategy for Neighbourhood Renewal'
Social Exclusion Unit (2000) 'Minority Ethnic Issues in Social Exclusion and Neighbourhood Renewal - A guide to the work of the Social Exclusion Unit and the Policy Action Teams so far'
Speak, S. and Graham, S. (2000), 'Service Not Included: Marginalised neighbourhoods, private service disinvestment, and compound social exclusion', *Environment and Planning A*, 1985-2001
STA City of Rome's Mobility Agency (2000) EURoPrice Technical Paper 2: Priority Policy Issues Report
Starrs, M. and Perrins, C. (1989) 'The Markets for Public Transport: The Poor and the Transport Disadvantaged,' *Transport Reviews*, 9, pp.59-74
Symbol (1997) 'New Jersey Parking Enforcement Officers take Symbol's Technology to the Streets'. Press Release
http://www.symbol.com/news/pressreleases/paradigm4.html
The Observer (2001) 'Race in Britain: the facts'. 15 July @
http://observer.guardian.co.uk/focus/story/0,6903,521949,00.html
The Observer (2002) 'Low-cost Brits take Europe by storm'. 11 August @
http://observer.guardian.co.uk/uk_news/story/0,6903,772611,00.html
The Observatory (2001) 'Poverty in Nottingham' @
http://www.nottinghamcity.gov.uk/coun/department/chief_execs/policy/povertyprofile/foreword.htm
The Prince's Trust (2001) 'Young People & Ethnicity'. Factsheet 10. February 2001
The Stationery Office (2001) 'Residents' opinions regarding speed enforcement and speed cameras: Select Committee on Transport, Local Government and the Regions Appendices to the Minutes of Evidence' @ http://www.parliament.the-stationery-office.co.uk/pa/cm200102/cmselect/cmtlgr/557/557ap12.htm
Townsend, P. (1979) *Poverty in the United Kingdom: A Survey of Household Resources and Standards of Living*. Allen Lane, London
TraC at the University of North London (2000) 'Social Exclusion and the Provision and Availability of Public Transport'. Department for the Environment, Transport and the Regions, London. July. http://www.mobility-unit.dft.gov.uk/socialex2/
Transport 2000 Trust (2003) *At the leading edge: a public transport good practice guide*. London
Transport for London (2003) 'Draft Strategy for London' @
http://www.london.gov.uk/approot/mayor/strategies/transport/pdf/2challenges.pdf
Transport Select Committee (2003) 'First Report – Urban Charging Schemes'
http://www.parliament.the-stationery-office.co.uk/pa/cm200203/cmselect/cmtran/390/39002.htm
Transport & Travel Research (2002) 'Measuring the Success of Marketing the Greater Nottingham Local Transport Plan'. Final Report to Nottingham Development Enterprise. Issue 1, January

Turner, J. and Pilling, A. (1999) 'Integrating Young People into Integrated Transport: A community-based approach to increase travel awareness'. Paper presented to a PRTC Conference on Young People and Transport, November 24th

Unison (2001) 'Car Parking and Public Transport Issues'. The Flyer, Nottingham Trent University @
http://science.ntu.ac.uk/unison/Flyers/2001/11.November%202001/Flyer%200 1.11.01.pdf

Victoria Transport Policy Institute (2003) 'Glossary' @
http://www.vtpi.org/tdm/tdm61.htm

Warren Centre for Advanced Engineering (2002). 'Transport Pricing: More than just a tax'. Sydney http://www.warren.usyd.edu.au/transport/publications.htm

Weinberg, A. and Ruano-Borbalan, S-C. (1993) 'Comprendre l'Exclusion'. *Sciences Humaines*. 28, 12-15

Wilson, D. (1999) 'Exploring the Limits of Public Participation in Local Government,' *Parliamentary Affairs* 52 (2) 246-259

World Bank (2001) 'Urban Transport Audits'. Infrastructure Notes Transport No. UT-2 @ http://www.worldbank.org/transport/publicat/td-ut2.htm

Index

For Product Safety Concerns and Information please contact our EU
representative GPSR@taylorandfrancis.com Taylor & Francis Verlag GmbH,
Kaufingerstraße 24, 80331 München, Germany

Printed and bound by CPI Group (UK) Ltd, Croydon, CR0 4YY
02/05/2025
01859336-0002